INTRODUCTORY OCEANOGRAPHY

INTRODUCTORY OCEANOGRAPHY

Joseph Weisberg
Associate Professor and Chairman
Department of Geoscience
Jersey City State College

Howard Parish
Assistant Professor
Department of Geoscience
Jersey City State College

McGraw-Hill Book Company

New York   St. Louis   San Francisco   Düsseldorf
Johannesburg   Kuala Lumpur   London
Mexico   Montreal   New Delhi
Panama   Rio de Janeiro   Singapore   Sydney   Toronto

Library of Congress Cataloging in Publication Data

Weisberg, Joseph S
  Introductory oceanography.

  1. Oceanography. I. Parish, Howard I., joint
author. II. Title.
GC16.W43   551.4'6   73-13890
ISBN 0-07-069046-4

Introductory Oceanography

1234567890MAMM79876543

This book was set in Musica by University Graphics, Inc. The editors
were Robert H. Summersgill, Nancy L. Marcus, and Barry Benjamin;
the designer was John Horton; and the production supervisor was
Bill Greenwood. The drawings were done by Vantage Art, Inc.
The printer and binder was The Maple Press Company.

CONTENTS

PREFACE

"Introductory Oceanography" originated with our experiences teaching and developing a newly created geoscience curriculum at Jersey City State College. One facet of the program involved an attempt to incorporate basic materials from the earth sciences in the "general studies" track. The result was the creation of a course in oceanography.

The oceanography course became one of the most popular courses offered, particularly among nonscience majors and especially those who are oriented toward the social sciences and education. In this course, basic concepts of the scientific endeavor can be taught using the students' natural interest in a "Jules Verne" world as a pivot.

Accordingly, our approach emphasized what we consider to be the basic investigations in the varied disciplines of modern oceanography. This text reflects what we believe to be an evenhanded treatment of physical, geological, and biological oceanography. The methods of investigation utilized by all three branches of oceanography are discussed, as well as a review of oceanography as an evolving science from the ancient world to the present era.

Most important, we have attempted to show how man hopes to utilize, to the fullest extent, the food and mineral potential of the sea; the controversy over this potential exploitation is not ignored. Throughout the text, in addition to investigative procedures and techniques, we have tried to show the effects of the sea and energy processes on the rest of our planet, and most especially the influence of the sea on our atmosphere. This view of the sea and the other parts of our planet acting as a single, unified whole is reemphasized and culminates in the last portion of the text.

We hope that this text will prove readable and understandable to the student. Most important, we trust that it will lead to a better understanding of the nature of our planet and its future; this we believe is essential for the so-called "literate citizenry" desired by all concerned science instructors if our planet is to have a future.

The authors must express gratitude and thanks to several individuals at McGraw-Hill who have pushed us forward at the right time and redirected us when necessary. We thank Bradford Bayne for his initial faith in the idea, Jack Farnsworth for his early planning of the book, Robert Summersgill for his final assistance in completing the text, and especially Nancy L. Marcus and Barry Benjamin for their continuous efforts and advice during all parts of the planning, writing, design, and execution of this work.

Joseph Weisberg
Howard Parish

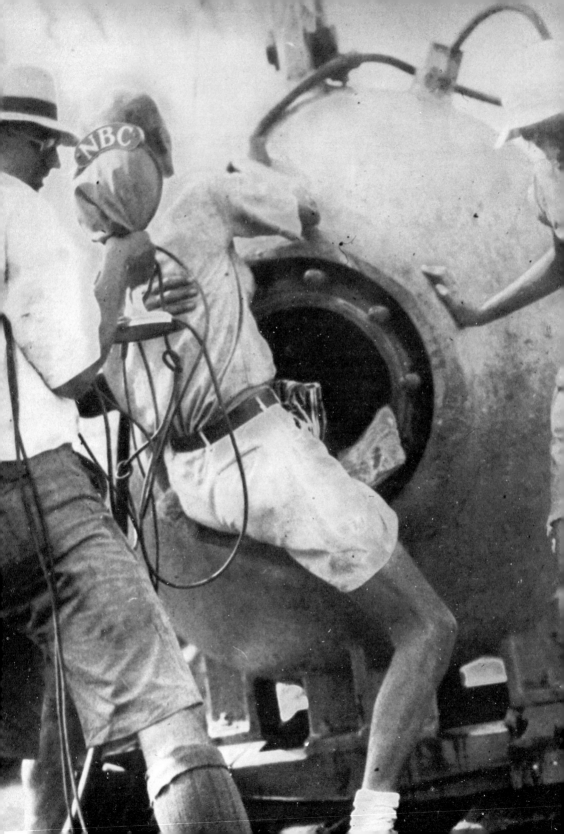

1 | STUDYING THE SEA

Early man viewed the sea as a forbidding, hostile, and bottomless abyss inhabited by the most frightening creatures imaginable. Even today, in our presumably informed society, the ancient sea monsters have been translated into mythical sea serpents like the Loch Ness monster. However, for the better part of our history the sea has served as an avenue of trade, transportation, and exploration. The seas—primarily the Indian Ocean and Mediterranean Sea—were traveled by mariners from many nations in a quest for treasure and new lands to conquer (see Fig. 1-1). As a result of these explorations, early advances in man's knowledge of the oceans primarily involved the geography of the seas and the continental boundaries. The concern with the geography of the oceans led to the use of the term *oceanography*, even though the word no longer describes the precise nature of our investigations today. Ocean geography was the primary purpose of investigations of the sea well into the nineteenth century. Our present modern oceanographic efforts and techniques result from the innovations made by a large number of researchers who lived during the last 100 years.

Although our study of the sea continues a number of studies begun even before the nineteenth century, our present approach to ocean exploration differs in several respects. Whereas the early explorers were searching for new trade routes, modern oceanography is a scientific study of all the physical, chemical, and biological aspects of the oceans. In fact, some scientists have suggested the alternate term *oceanology* for the present-day study of the

William Beebe entering the bathysphere for one of his early descents into the ocean.

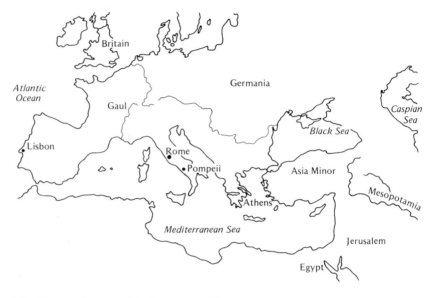

1-1   The ancient world about A.D. 150.

sea; the word literally means ocean study. Oceanology encompasses the total data collected by specialists from all scientific disciplines. This multidisciplinary approach to the examination of the sea attempts to develop an understanding of the sea in relationship to the Earth's total environment.

Oceanography in its modern role as the science of studying the world's oceans utilizes the services of all branches of science. In the examination of the seas we see specialists contributing their special knowledge from the geological, meteorological, physical, chemical, and biological disciplines (see Fig. 1-2).

The data derived from ocean exploration are examined from a number of interconnected scientific viewpoints. Water is investigated for its chemical composition and its circulation. Biologists attempt to examine the organisms in the sea in relation to the chemical composition of the surrounding water and the effects of the physical factors of heat, light, and pressure. Physicists are concerned not only with the physical characteristics of water resulting from its composition and the effects of depth but also with the nature and movements of the rocks and sediments of the sea floor. In this endeavor the geologists, too, are attempting to learn more about the sea floor and the Earth's interior. Thus, the modern studies of the sea are the extensions of examinations that were begun on the continents.

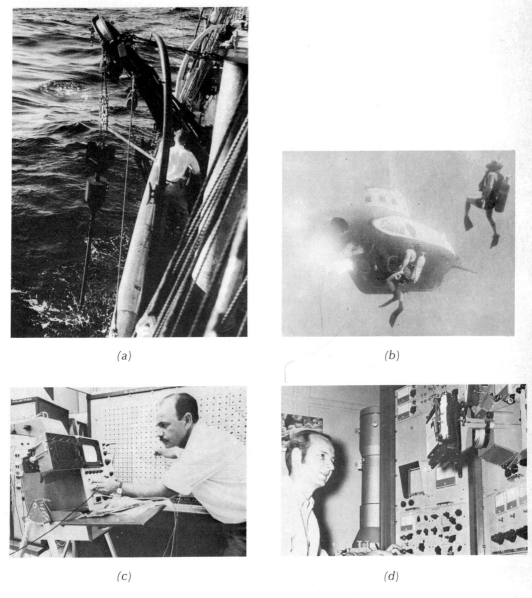

(a)

(b)

(c)

(d)

1-2 The science of oceanography depends upon the talents of scientists from many disciplines. Cooperative ventures utilize both shipboard and laboratory investigations. *(a)* Piston *corer* being recovered from R.V. *Vema*. *(Lamont-Doherty Geological Observatory)*
(b) Research submarine *Star III* with divers. *(General Dynamics)*
(c) Chemical analysis of water samples. *(UN-FAO photograph)*
(d) Electron-microscope analysis of specimens. (USDA)

THE ROOTS OF OCEANOGRAPHY

Modern oceanographic procedures have been attributed to many individuals, but the investigations that are now conducted by scientists are the result of long-term accumulations of data. The development of the numerous techniques of data collecting and recording have resulted from man's natural curiosity about the planet Earth from ancient times to the present.

The ancient sailors contributed primarily to the discovery of new trade routes. As a result of voyages beyond the Mediterranean Sea, the boundaries of the known world were gradually extended (see Fig. 1-3). These explorations led to information upon which further examinations were based. Early voyages by the Minoans of Crete, the Phoenicians, and later the Romans (Fig. 1-4) saw the known world

1-3 The major trade routes of the ancient world. *(After ARAMCO Handbook)*

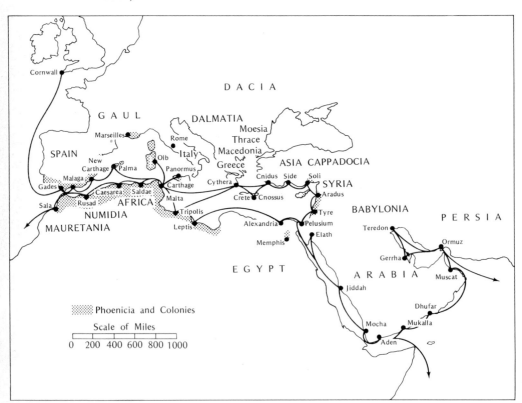

1-4   A Roman galley.

extended from the Mediterranean Sea around Africa to the British Isles even before the birth of Christ.

These voyagers often sailed on boats made of papyrus, wood, and other materials that frequently were unseaworthy for long and extended trips. Furthermore, navigation was difficult due to the lack of instruments and the inaccurate star charts used to determine position. As a result, most early voyagers traveled close to the coastline and utilized land sightings to determine their course, only rarely venturing into the open sea.

Gradually, more seaworthy craft and better navigational devices were developed through the contributions of many individuals from several seagoing civilizations. As the known world expanded, the nature of the oceans and their differing characteristics became more apparent. The presence of ocean currents was recognized by early explorers, but the pattern of the shifting currents remained a puzzling phenomenon. The fact that winds and currents form a worldwide pattern and consistent scheme that is coherent and orderly went unrealized until the middle of the nineteenth century.

Data describing various oceanic phenomena were recorded and carried forward from the ancient literature. The Greeks, who attempted to classify all living things into general groups such as land dwellers, air dwellers, and water dwellers, recognized variations even within these categories. The Greeks were the first to recognize that some sea dwellers such as whales, dolphins, and porpoises are mammals and biologically quite different from the fish that share their water environment.

The Greeks were a curious, logical people who laid down a vast literature covering many branches of science. Although they established the tradition of recorded data in scientific observations, mysticism and superstition became intermingled with science, and man's world view was affected by a number of erroneous constraints. With several notable exceptions, the Greeks rejected the view of a round Earth revolving about the sun. They envisioned our planet as a stationary, flat disk encircling the Mediterranean Sea. Beyond this small world lay the Atlantic Ocean, which was, in turn, bounded by a vast whirlpool ready to engulf mariners who might venture to its rim.

Gradually, the known world was expanded, as a consequence of the later voyages of the Irish and the Vikings, who ventured into the North Atlantic Ocean. By A.D. 1000, Greenland, Iceland, and other parts of the New World had been explored by these people. Many small colonies were established in these new lands, but they were short-lived. During the Age of Discovery, a period of time that marks the era of the numerous European voyagers, the true extent of almost all the ocean boundaries was properly defined. Contributions were made by explorers such as Bartholomeu Diaz, Vasco da Gama, Balboa, Columbus, and Magellan (see Fig. 1-5).

Although many of these voyages, and those to follow, were made primarily for conquest and trade, much information was contributed to the general ocean lore that eventually became part of our scientific literature. Between 1769 and 1779, Captain James Cook (Fig. 1-6), the discoverer of Australia, undertook three voyages in an examina-

1-5   Magellan's voyage of discovery.

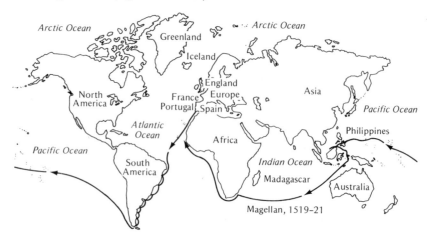

1-6   Captain James Cook, R.N. *(U.S. Navy)*

tion of the world's oceans. During his earliest voyage he noted the general configuration of the Pacific Ocean.

Although Cook was also seeking the legendary southern continent, Terra Australis, he never sighted the landmass. However, his observations of numerous seabirds and the presence of icebergs allowed him to infer the existence of the southernmost continents as the origin of the birds and ice floes. Therefore, Cook is credited with the discovery of Antarctica, as we now call the landmass, despite the fact that the continent was first sighted by Charles Wilkes, an American, in 1840. Thus, Cook's deduction of the existence of Antarctica, although it required later confirmation by Wilkes, is accepted as an accomplished fact. This technique of utilizing inference from scattered data to draw a conclusion is frequently utilized in the beginning stages of a scientific investigation.

Cook made a third voyage into the northern seas in search of the Northwest passage, a sea-lane which, it was believed, offered a shorter northern route from Europe to the Pacific. Cook was unsuccessful because of the impossible ice-covered waters of the Arctic Ocean. He then returned to the North Pacific for further exploration of Hawaii, where he was killed by the inhabitants.

The Northwest passage was finally discovered by Roald Amundsen during 1903–1906. He, too, found the region impassable, and the region was not opened as a new trade route until very recently (see the S.S. *Manhattan*, p. 19). Amundsen was the first man to reach the South Pole, a feat he accomplished in 1912.

MODERN OCEANOGRAPHIC TRADITIONS

One of the early American pioneers in oceanographic investigations was the versatile Benjamin Franklin (Fig. 1-7). Although Franklin held many diplomatic and political posts and was a writer and printer, during his own lifetime he was perhaps more famous in Europe as an inventor and scientific investigator.

Franklin used ships' logs as the source of data on the action of a huge current of water now known as the *Gulf Stream* (Fig. 1-8). Franklin's description of this immense flow of water allowed traders to avoid its eastward flow when they traveled from Europe to the Americas. Thus, the colonists' ships saved several days each trip as they plied their trade between the Old World and the New. This charting of the Gulf Stream by Franklin in 1786 was most important to the developing economy of our young nation. Millions of dollars in shipping costs were saved each year during the early days of our country's development. This technique of data synthesis from diverse sources is one that would play an important role in later oceanographic studies in physical oceanography during the middle of the nineteenth century.

Franklin also made several voyages between the two continents. During these voyages, he recorded temperature readings of the surface water in the Atlantic Ocean. For these measurements he used a thermometer placed in a bucket of water collected from the surface.

Just as Franklin was noted for having an interest in a wide variety

1-7   Benjamin Franklin. *(Division of Graphic Arts, Smithsonian Institution)*

of scientific disciplines, a contemporary of Franklin's also exhibited a similar wide-ranging interest in the science of our environment. Alexander von Humboldt (1769–1859) was a naturalist who as a relatively young man had inherited a large sum of money and was thus able to follow his interests as he saw fit. During 1799–1804, Humboldt sponsored an expedition to South America to study the characteristics of the continent.

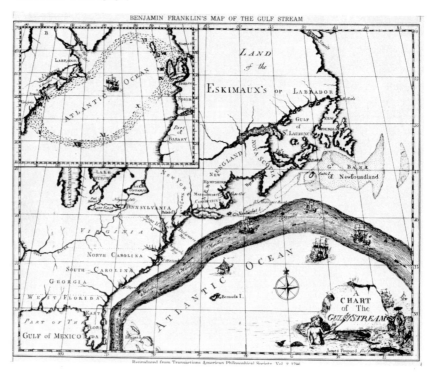

1-8 The map of the Gulf Stream developed by Benjamin Franklin. *(U.S. Naval Oceanographic Office)*

The examinations and explorations carried out by Humboldt during this time were immense. He explored the flora and fauna in the areas he visited, examined the soil, studied the meteorology, investigated earthquakes, and studied astronomical objects. Humboldt also noted the cool, foggy climate of Peru, which results from the presence of the Peru Current. This current, originally called the *Humboldt Current,* moves south to north along the west coast of South America. Here, prevailing winds cause cool bottom water to move upward in a process called *upwelling* (see Chap. 8).

Humboldt attempted to examine the varied aspects of nature as a single, unified structure. His unique view of nature in its entirety required Humboldt to analyze vast amounts of data from a number of viewpoints. He attempted to snythesize diverse data into an ordered scheme. His 40,000-mile journey took 30 years to analyze, and the final report filled 30 volumes of data and conclusions.

THE FOUNDERS OF MODERN OCEANOGRAPHY

Modern oceanography may be traced to Matthew Fontaine Maury (1806–1873), a lieutenant in the U.S. Navy (Fig. 1-9). Maury's main concern was an examination of the sea and its physical characteristics. In 1836, he published his first text on the subject of navigation and 3 years later began the work that was to win him world fame.

In 1839, Maury began an intensive investigation of ships' logs. Like Franklin, Maury examined the sailing times, winds, currents, and other factors which affected the movement of the ships. In 1842, Maury was put in charge of the U.S. Naval Depot of Charts and Instruments in Washington, D.C. This appointment was probably one of

1-9 Matthew Fontaine Maury. *(U.S. Naval Observatory)*

the most fortunate ever made by the United States government. As officer in charge, Maury had at his disposal vast amounts of data about currents and winds recorded in ships' logs collected from all parts of the world. Many seafaring nations cooperated in the undertaking. The logs were sent to the National Observatory in Washington, where they became available to Maury. Thus, he was able to correlate and chart the work of thousands of people who were making records in all the major oceans of the world. These charts of currents compiled by Maury enabled ships' captains to cut 50 days off the London to San Francisco run, thereby saving countless millions of dollars.

In 1855, Maury's work resulted in "The Physical Geography of the Sea," the world's first English text dealing with the science of physical oceanography. The book dealt with surface currents in the open sea, depth readings in the ocean, and the character of the ocean basins.

In 1853, Maury initiated and planned the first international science conference in oceanography. Held at Brussels, it was designed to stimulate interest in establishing a marine weather-observation sys-

1-10   Comparison of fathoms, feet, and meters.

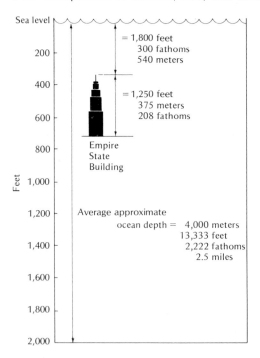

Sea level

200

400

600

800

Feet   1,000

1,200

1,400

1,600

1,800

2,000

= 1,800 feet
300 fathoms
540 meters

= 1,250 feet
375 meters
208 fathoms

Empire
State
Building

Average approximate
ocean depth =   4,000 meters
13,333 feet
2,222 fathoms
2.5 miles

tem to bring order out of the countless chaotic observations then being made. Biologists, too, were becoming part of this evolving science of the sea. Edward Forbes (1815–1854), born on the Isle of Man, was one of the first internationally known biologists to develop theories about ocean life. Forbes was a naturalist who examined the geology and biology of the marine life-forms he found as fossil remains. He took special note of the bottom configurations and how they affected life-forms. Forbes's chief contribution was as the originator of the science of marine biology. In addition, his work inspired others to seek knowledge about marine life and helped to advance the science.

Forbes is probably best remembered for his most erroneous theory, that of the Azoic zone. In 1850, he devised a theory which divided the ocean life-forms into several habitable zones and one zone, called the Azoic zone, assumed to have no life. This zone existed below 300 fathoms, or 1,800 feet (see Fig. 1-10). It was believed that in this zone, near the bottom, oceanographers would find the primal ooze which gave rise to all life at the beginning of time. The ooze was named *Bathybius* by Thomas Huxley, the foremost advocate of Darwin's theories in England.

APPROACHES TO STUDYING THE SEA

Modern studies of the sea utilize a number of techniques and devices. However, the examinations are carried out by different approaches; the manner of data collecting varies with the nature and extent of the region to be studied. These approaches and innovations were devised by numerous investigators during several outstanding voyages undertaken over the last 100 years. These voyages differ from earlier studies in that their researchers utilized and devised instruments to measure specific characteristics of seawater and the ocean basins. Specimens of marine life, ocean water, and sea-floor samples were collected for various detailed analyses. Oceanography began to move toward a quantitative base from an original mixture of descriptive sea lore and scattered research.

THE VOYAGE OF THE *CHALLENGER*
The ideas developed by Forbes and the concept of *Bathybius* were part of the impetus that led to the most important oceanographic voyage of all time. This journey was the route followed from December 7, 1872, to May 24, 1876, by the British ship *Challenger*, a 2,306-ton corvette with auxiliary steam power. The ship was the first to be

completely outfitted for large-scale oceanographic research (see Fig. 1-11).

The *Challenger* expedition was headed by Charles Wyville Thompson (1830–1882), a student of Forbes, assisted by John Murray (1841–1914). After the death of Thompson, Murray continued the analysis of the data gathered by the expedition.

The *Challenger* crisscrossed the North Atlantic, traveled into the South Atlantic, voyaged into the Antarctic Ocean, visited Australia and Hawaii, sailed through the Straits of Magellan, and returned to England in 3½ years. The 68,890 nautical miles covered during this journey still ranks as the longest continuous scientific expedition in history. (A nautical mile is 1.18 statute, or land, miles.)

During the journey, maps of the ocean bottom were made in many places. Both surface and deep-water temperatures were recorded in the oceans visited by the ship. The *Challenger* was equipped to make magnetic, meteorological, and hydrographic measurements. In addition, numerous samples were collected by means of water bottles, dredges, and other devices. In the Mariana Islands, near the coast of Guam, the *Challenger* made a depth sounding of 26,850 feet, an astounding record at that time.

All the oceans were examined for animal and plant life. A total of 4,717 new specimens were collected, consisting of macroscopic and

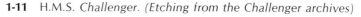

1-11   H.M.S. *Challenger. (Etching from the Challenger archives)*

H. M. S. Challenger

microscopic life-forms. Furthermore, life was found at all depths of the ocean. The theory of *Bathybius* was overthrown, for no primal ooze was ever found during the voyage (and, of course, never has been found). The sediments recovered from the ocean floor were of different origins; some had formed in the ocean, and others had been transported from the land.

At the end of the voyage, the collected information resulted in a comprehensive report and analysis. The completed record consisted of 50 volumes and required almost a full generation to compile!

The *Challenger* inspired a great number of research voyages throughout the rest of the nineteenth century. However, the *Challenger* expedition still reigns as a supreme example of a type of study known as individual interrupted observations, a large-scale study of an immense area by a single vessel.

THE CHANGING CHARACTER OF OCEANOGRAPHY

One of the leaders in oceanographic research during the latter part of the nineteenth century was Alexander Agassiz (1835–1910), son of Louis Agassiz, the great Swiss naturalist who taught at Harvard University. Alexander was a successful businessman whose fortune allowed him to outfit and pursue oceanographic research. From 1877 to 1905, he covered 100,000 miles and developed several new tools and techniques for direct measurement of ocean-bottom topography. He was the first to develop a mechanical steel-cable system for depth soundings; he invented a new type of trawl for dredging; he also studied the life in the Caribbean Sea, mapped Pacific topography, traced the circulation of the Gulf Stream, examined populations of microorganisms such as plankton, and noted the similarities between the Caribbean Sea and the Pacific Ocean.

By the turn of the century, many voyages were made by Norwegian, Dutch, English, French, American, Japanese, and many other seafarers. One of the most unusual and outstanding men of the period was Albert I, Prince of Monaco (1848–1922), the ruler of Monaco from 1889 to 1922.

The prince was a naturalist and oceanographer who used several of his own yachts for oceanographic research. He tracked drifting floats in currents of the Atlantic Ocean and Mediterranean Sea. He became a world-recognized expert on the currents of the North Atlantic and the Mediterranean and studied the topography of the Atlantic Ocean floor.

The prince was one of the first men to study the giant squid and its habitat. During his lifetime, he founded the Oceanographic Museum and Laboratory in Monaco and established, as a gift to the French people, the Oceanographic Institute in Paris. These research institutions have become important world centers for the collection of data and have made France and Monaco world-renowned in oceanographic research.

Many other voyages and expeditions set sail during the twentieth century. Eventually these expeditions grew into international cooperative undertakings on an increasingly large scale.

RECENT OCEANOGRAPHIC EXPEDITIONS

One of the most famous oceanographic expeditions undertaken by a single ship was the voyage of the *Meteor* during 1925–1927. This voyage was the first systematic study of a single ocean ever undertaken by an oceanographic research team. The German ship crossed and recrossed the Atlantic Ocean a dozen times during its 2-year voyage, gathering weather reports, water samples, and information about the irregularities of the ocean floor. Electronic apparatus allowed the *Meteor* research team to make continuous records of the ocean-floor depth rather than selected measurements, as in the past. Later, this information was not only to be of scientific value but became extremely important to the German mariners during the submarine warfare of World War II. The voyage of the *Meteor* still remains as a classic example of continuous observations made in a single ocean study.

The IGY

The *Meteor* was the only complete survey of a single ocean study until the International Geophysical Year (IGY) of 1957–1958. This was a year of great international cooperation in all the geosciences, undertaken by 37 nations which supplied a total of 80 ships for oceanographic research. During this research venture, simultaneous observations of selected portions of the world were made by several ships and their complement of instruments. In different regions, other vessels carried on continuous observations of physical characteristics.

During the IGY, one British and three American ships crisscrossed the Atlantic Ocean from north to south in a wide-ranging, systematic study of the entire ocean. In addition to measurements of the topography, the studies centered on magnetic surveys, charting of ice floes, meteorology, atmospheric circulation, currents in the ocean, carbon dioxide concentrations in the air and water, and testing new devices and equipment.

IGY scientists examined the entire Atlantic Ocean and most of the

Pacific Ocean. In addition, the first extensive study of the Indian Ocean was carried out. Although the Indian Ocean was one of the first oceans to be utilized as an avenue of trade and exploration, the physical characteristics of this ocean were the last to be studied.

Some of the data gathered during the year included the dating of Antarctic water, which revealed that portions of the bottom water may be as old as 2,000 years. (How the "age" of water is determined is discussed in Chap. 2.) Samples of glacial ice were taken and

1-12 Manganese nodules dredged from the sea floor.

analyzed, showing that there has been a warming trend in the atmosphere of the Earth which may be as much as 2 to 3°F every 100 years. Manganese nodules, a possible source of manganese and other metals, were discovered in great abundance on the ocean floor, leading to the development of techniques for their recovery in large quantities at some future date, (Fig. 1-12). Finally, many ridges were noted and mapped. The IGY revealed a uniformity in the geology of the world's oceans which had not been fully realized before.

The IGY also led to other special study committees in the hope that the great cooperation shown during the year could be maintained for other research on a continuing basis. These committees established other investigations, some of which are concerned primarily with the Indian Ocean and the floor of the Pacific Ocean. These continuing studies of the oceans and the atmosphere mark the latest phase in our developing examinations of the planet Earth. Studies of the sea now include concurrent examinations of the atmosphere in attempts to define the relationships between the energy exchanges and composition of these two media, air and seawater. Large-scale international cooperation is now the major concern of oceanography and its sister science, meteorology.

SCOR

Following the IGY, the Special Committee on Oceanographic Research (SCOR) continued the studies carried on during the IGY. SCOR is a special committee which was established as a cooperative undertaking of England, India, Russia, and the United States under the auspices of the International Geophysical Union. Its function is to coordinate and initiate research in oceanography. Since IGY it has placed a great emphasis on projects in the region of the Indian Ocean in addition to its other activities.

Today, oceanographic information from all the world's oceans is exchanged among nations and international societies on a daily basis. On research vessels one can find scientists of many nations contributing their special knowledge to the success of the investigations. Data recovered from the sea floor are investigated by Americans, Russians, Frenchmen, Englishmen, and other nationalities. Cooperative undertakings are under way by special agreement between countries and through the United Nations.

Many more agreements are needed if we are to avoid the waste and competition which marked past periods of exploration. But our experience has yielded, thus far, much more of a cooperative effort in oceanographic and meteorological studies than ever before.

1-13 S.S. *Manhattan,* a supertanker-ice breaker. [*Standard Oil Company (N.J.)*]

SUMMARY QUESTIONS

1. It is often said that oceanography is a misleading term in that there is no one study that can accurately be called oceanography. Explain why this is a true statement.

2. What factors largely forestalled a worldwide ocean exploration prior to the advent of the Vikings in the North Atlantic?

3. In what way were the inferences made by Franklin and Maury similar? How did their data gathering compare?

4. Describe the three major types of oceanographic exploration studies. In what ways are they similar? How do they differ?

5. Describe the nature of the work carried on by H.M.S. *Challenger* and some of the significant discoveries it made.

6. In what significant way was the IGY a forerunner of the type of studies that will be done in the future? Describe the kind of continuing work that led from the IGY initial studies.

7. List the various experimental and investigative approaches that make the oceans a place for much interdisciplinary study.

8. What was the primary goal during the Age of Discovery?

9. How does the modern oceanographic approach to studying the sea differ from the concerns and approaches before the eighteenth century?

10. How was Captain Cook credited with the discovery of Antarctica even though he never actually sighted the landmass? What important scientific process did this "discovery" utilize?

11. What important practical results accrued almost immediately from the work done by Maury and Franklin?

12. How do the *Meteor* examinations compare with other types of oceanographic examinations? What significant data were developed by this research ship and team?

13. For what reasons were most ancient voyages carried on close to land?

14. What were some of the important data gathered during IGY? Include several different areas of investigation.

15. What was *Bathybius?* How did this concept, in a very peculiar way, contribute to the advance of oceanography?

16. How many fathoms did the *Challenger* record in the Mariana Trench? *Hint:* See Appendix Table A-6 for conversion factors.

17. What is the trend on earth in regard to climatic changes? How might this effect the oceans?

18. Although Forbes was incorrect in some of his theories, why might it be said that errors, too, help to advance scientific progress?

19. Why is the voyage of the *Challenger* often considered to be the beginning of modern oceanography?

20. Explain the work done by at least three different scientific disciplines into oceanographic research.

21. Why are the Greeks so often credited with founding the nature of the scientific enterprise?

22. What was the significance of the *Meteor* expedition?

23. Why are oceanography and meteorology considered "sister sciences"?

24. Why are current charts helpful in saving costs of shipping freight?

2 | MEASURING THE PROPERTIES OF THE SEA

As oceanographic investigations progress from a limited series of examinations to more detailed sets of data, greater accuracy of measurement becomes more important. The measurement of the various physical properties of the sea, and of the Earth itself, is still carried out primarily from the surface. Oceanographers are largely still bound to working from the decks of surface research vessels. Thus, the development of instruments has centered on techniques that can be used to produce measurements by indirect processes. This is accomplished by devices that make recordings and collect samples from great depths while being controlled from the surface.

Of course, man has always been concerned with developing techniques of diving and the improvement of submersible research craft. An ancient legend says that Alexander the Great descended into the Mediterranean Sea in a primitive diving bell (Fig. 2-1). It was shaped like an inverted glass so that as it was lowered, air remained trapped inside. Upon being lowered into the sea, the diver in the bell made a series of observations of the area surrounding the bell. One reported incident tells of huge fish, which flashed by Alexander in his submerged bell at a fantastic speed—faster than the eye was able to follow. The fish was so huge that it required several days to pass by the diving bell.

Although the story of the diving bell and the fish is apocryphal—to say the least—it does reveal man's great desire to find what secrets lie below the first few feet of the ocean surface. Firsthand observations of the sea floor have always been one of man's cherished dreams.

R.V. *Chain*—the major research vessel at Woods Hole Oceanographic Institution. (With permission of Woods Hole Oceanographic Institution)

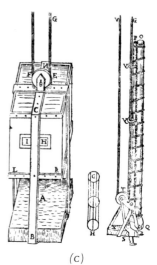

(a)   (b)   (c)

2-1   Ancient diving equipment. *(a)* Model of an early "do-it-yourself" diving apparatus; *(b)* earliest known diving suit; *(c)* diving bell by B. Lorini. *(The Smithsonian Institution)*

Only recently have men begun to descend into the murky depths of the world's oceans. Special craft and instruments now carry man and his recording devices to almost every part of the ocean floor (see Fig. 2-2), and although we have not discovered anything as fantastic as Alexander's fish, many fascinating discoveries are made.

The major methods of ocean-floor examinations and measurement have been by indirect means and continue to be at the present time. Specimens of rock, water, and living organisms still are obtained chiefly by a variety of devices operated from the decks of ships. Measurements of chemical and physical properties of water and geology of ocean basins are made by complex scientific instruments.

The investigation of ocean water, for example, involves a series of examinations of its physical and chemical characteristics. These tests are made both in the field and from specimens gathered from a variety of depths in order to determine the chemical characteristics of the water. Water samples must be collected so that salinity (dissolved salts) and other chemical properties can be analyzed in the laboratory. Chemical composition as well as temperature varies from one locale to another, and the changes also occur with depth. These determinations, and others, must be made both in the laboratory and on the site by special devices and techniques.

2-2   A bottom current meter is put over the side. *(NOAA)*

METHODS OF MEASUREMENT

Even in a small area, the general character of the ocean and its physical properties vary greatly. As a result, many observations from widespread sources must be obtained and analyzed. Often, these examinations are *simultaneous readings,* such as the large amount of data recovered from all vessels involved in the research during the IGY of 1957–1958. In other situations, the data are gathered during a *continuous study* of a region, such as the SCOR data from the Indian Ocean.

Voyages such as those of the *Meteor* and the *Challenger* are supreme examples of methods of data collecting from a *single source* covering a wide area. This information is then correlated with other data taken by other single sources in the same area obtained at different times.

RADIOACTIVE CARBON DATING

In order to "date" ocean water, the quantity of the naturally radioactive carbon 14 is measured. This form, or *isotope,* of carbon becomes part of the carbon dioxide in the atmosphere and enters the ocean at the surface. The C^{14} isotope decays with a half-life of 5,600 years; that is, in 5,600 years one-half the original will remain. In 11,200 years the C^{14} will decay to one-fourth the original (one-half of one-half), and the process of decay continues with an accuracy of measurement of about 50,000 years.

Deep water is analyzed for the quantity of C^{14} present. The amount of C^{14} compared to the common C^{12} allows one to date the period when the deep water was last in contact with the surface and the atmosphere.

Three of the most important factors which are investigated most intensely are the temperature, density, and salinity of the water. Other determinations reveal more specific information.

For example, an examination of the oxygen concentration of bottom water shows that the gas concentration decreases with depth to a limited extent. The decrease is related to the length of time the water has been in that region. Thus, oxygen concentration may be used to approximate the "age" of ocean water and to indicate past climatic factors which affected its development.

Carbon 14 radioactivity determination has been used to calculate the age of water, that is, the frequency of mixing and the length of time the water mass has spent beneath the surface of the water body. From the amount of the radioactive isotope present in the water we know that the age of water masses varies widely, even within small areas. It has been shown that the general circulation and renewal of large masses of deep water occurs very slowly. For example, it has been determined that the North Atlantic central water is about 600 years old, while the bottom water is almost 900 years old. The Antarctic central water and South Atlantic bottom water are both approximately 300 years old, but the deep South Pacific water is the oldest of all, somewhere between 1,100 and 1,900 years old.

Conversely, examinations of the surface waters reveal that they mix and overturn at a fairly rapid rate. In recent years, fallout of radioactive particles from atom-bomb experiments have been collecting in large quantities in the ocean. Examinations of the movement of these materials reveal a much more rapid mixing of the upper surface waters of the ocean than was previously assumed.

PROFILE GRAPHS

In an attempt to correlate as many diverse data as possible, oceanographers draw profile graphs for data such as temperature, salinity, and density as they are related to depth. The profiles produce a schematic diagram for specific sets of data on a vertical scale within a prescribed region.

Profiles such as these readily reveal, on a visual scale, the geographic distribution of the data over great distances. Since the various characteristics are dependent upon one another, the diagrams reveal the various vertical, geographic, and horizontal fluctuations when each factor is plotted against the others. An analysis of such data shows that each water mass within the ocean has a specific combination of physical characteristics. The profiles that are developed depend upon the accuracy of measurement of each of the physical characteristics (see Fig. 2-3).

Density

Density determinations, that is, the mass of a water sample per unit of volume, are the most difficult to obtain directly in the field. The

---

2-3 A temperature-salinity diagram showing the changes in the characteristics of seawater produced by variations in temperature and salinity.

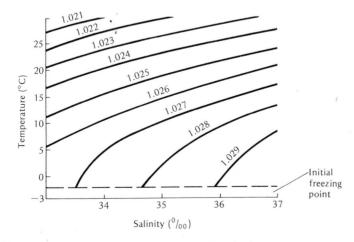

techniques and instruments are not yet sufficiently perfected to allow fast and accurate direct measurements.

Rapid measurements are made easily by means of a hydrometer. This device is a weighted tube which floats at different levels in the water sample, the level being determined by the density of the water. But since density is determined partially by temperature for example, removing the sample from its source will affect the reading.

Most often, density determinations are made indirectly, being computed from the temperature and salinity measurements calculated by specific techniques. At this time, accurate, fast methods for shipboard density determinations have not been developed.

Salinity

Salinity measurements can be obtained by chemical techniques. In the process known as *chlorinity titration* chemical indicators are used in either a physical examination or electronic analysis.

Accurate salinity examinations, however, are usually made by measuring the electric conductivity of the water. Pure water is not an electric conductor, but the salts dissolved in seawater carry electric currents. Thus, the amount of current which can pass through a sample of water is a direct result of the number and quantity of dissolved salts in the water.

Ships use a device known as a *toroid cell* to measure the conductivity, and thus the salinity, of ocean water. The unit consists of a series of wire-wound cells. A current is sent through the wire when the cells are in the water. The voltage produced in the cells is in direct proportion to the salinity of the water surrounding the cells. The voltage is measured and used to determine the salinity of the water.

Temperature

A wide variety of instruments have been developed to measure the temperature of the water at various depths in the ocean. Some instruments produce quick readings in a single area; others give continuous large-scale readings covering a wide area of the ocean.

Thermometers   Surface readings are accurately and quickly taken by a simple thermometer. Thermometers that determine maximum and minimum temperatures are used to find the range in a specific area in shallow water. However, for accurate temperature readings at depth, *reversing thermometers* are used (Fig. 2-4).

Reversing thermometers are specially devised for depth readings. The thermometer has a fine capillary tube drawn out into a loop and a constriction just above the reservoir of mercury. When the ther-

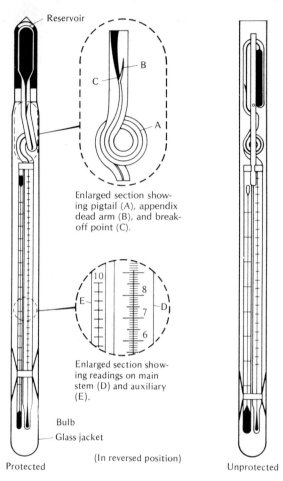

Enlarged section show-
ing pigtail (A), appendix
dead arm (B), and break-
off point (C).

Enlarged section show-
ing readings on main
stem (D) and auxiliary
(E).

Bulb

Glass jacket

(In reversed position)

Protected

Unprotected

2-4 Reversing thermometer. *(Modified from USNOO drawings)*

mometer is lowered into the water, the mercury can pass out of the reservoir into the looped capillary tube. At a predetermined depth, or when the thermometer is attached to a water-sample bottle, the thermometer is tripped over. This causes the mercury column to separate from the reservoir, thus trapping the mercury at the proper reading. When the thermometer is returned to the surface, the reading taken at the desired depth does not change.

In fact, reversing thermometers may also be used to determine depth. It is known that pressure affects the temperature reading on a thermometer. The pressure causes the mercury to produce a false reading, one that is higher than normal. Some water-sample bottles have two reversing thermometers. One thermometer is unprotected

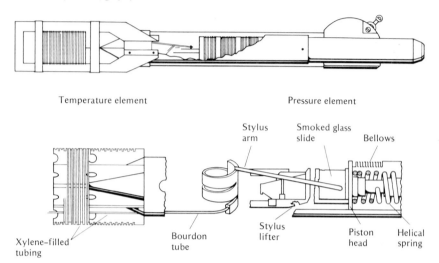

Temperature element   Pressure element

Stylus   Smoked glass
arm   slide   Bellows

Xylene–filled   Bourdon   Stylus   Piston   Helical
tubing   tube   lifter   head   spring

2-5   Internal mechanism of a bathythermograph. *(Modified from USNOO)*

and gives the reading as it is affected by pressure. The other thermometer is placed in a glass tube to protect it from the pressure of
the surrounding water. When the two readings are fixed by tripping,
a comparison between the two gives the oceanographer a means of
calculating the depth at which the water sample was taken.

The Bathythermograph   To determine water temperatures on a
continuous basis, oceanographers employ the *bathythermograph,*
which utilizes a temperature-sensitive liquid in a sealed container.
As the bathythermograph is lowered into the water, the liquid re-
sponds as the temperature changes with increased depth. The liquid
causes a stylus to move across a smoked-glass plate (Fig. 2-5). The
marked plate then becomes a permanent record of the temperature
changes with depth in the ocean. Accurate readings of this sort have
been made to depths in excess of 1,000 feet.

DETERMINING WATER MOVEMENT AND CHARACTER

Oceanographers are engaged in a continuous examination of the
ocean surfaces and depths to determine the characteristics of the
water and its motions. The physical nature of the ocean's waters and
their various interactions are under intense investigation to determine
the effect these factors have on one another, on the marine organisms

in the water, and ultimately on world weather and climate. Thus, if we are to understand the varied nature of our planet, we must begin with the ocean and its physical characteristics.

Ocean currents, large rivers of water within the sea, move along at all depths. Surface currents, as well as bottom currents, effect a mixing of the various physical properties of the ocean water. Many of these huge tongues of water are larger than all the world's land rivers combined, and vast quantities of water are carried by these ocean currents throughout all parts of the ocean. Tracing and measuring these rivers of water forms a major part of the physical oceanographer's work.

Current Meters

The speed and direction of ocean currents are determined by a variety of techniques. One of the most common methods is to use current meters. The basic type of current meter commonly in use consists of a rotorlike device which is turned by the current of water rushing past the anchored meter. As the rotor turns, it creates a small electric current in the meter. The speed of the rotor determines the amount of electric current produced, which is in direct proportion to the speed of the water in the ocean current. The speed is then transmitted through a cable to the research vessel, or a small radio transmitter sends the information back to the ship.

In addition to the rotor, a small compass also is included in the current meter. Thus, with one reading the operator is given the velocity of the ocean current and its direction as well.

Although current meters are relatively accurate, they do present some problems for the researcher. As the current meter moves up and down in the waves of the open sea, this motion affects the rotor and thus the accuracy of the speed determination. For this reason other devices are used for various types of readings.

Other Determinations of Currents

Often, only the average speed and direction of the current movements are needed since, in some investigations, only the general configuration of the current is necessary. When accuracy can be sacrificed for convenience, *drift bottles* are used. Drift bottles are sealed, weighted containers set to float at some predetermined depth. Inside the bottle is a card requesting the finder to fill in the place of recovery and mail the card back to the research institution. A scientific message in a bottle! The information obtained allows scientists to estimate the average speed and direction of the current.

Drift bottles also may be set to float at the surface. In this way they can be tracked directly and their direction and speed noted. Drift bottles are satisfactory for calculating the average speed and direction of a current. They differ from current meters in that they do not determine the many fluctuations that occur in the movement of the water.

Sound waves are also used to determine the speed and direction at which the ocean currents move by noting the difference in the velocity of two sounds. One sound wave is transmitted in the same direction as the current and compared to the velocity of a sound wave transmitted in the opposite direction (see Fig. 2-6).

Still another device utilizes two sensing probes, or electrodes, towed behind a ship. As seawater moves between the electrodes, an average velocity can be obtained as the water and electrodes cut the Earth's magnetic field, producing a small electric current. The electric current induced by cutting a magnetic field is called the *dynamo effect,* as it is similar to the way electricity is produced for home and industry. The voltage produced by the action of the moving water is in direct proportion to the velocity of the current.

Currents can also be traced for average values by radioactive dyes and tracers. The dyes can be observed and their movements measured; the radioactive materials can be tracked through the water by instruments which record the radiation released, and then the direction and rate of movement of the water are determined.

For current determinations below the surface, the most common device used today is the *Swallow float,* which is set to travel at a predetermined submerged depth. As the current carries it along, the float sends out a signal from a tiny sonic device within. The float is

---

2-6 Sound signals used to measure ocean current flow.

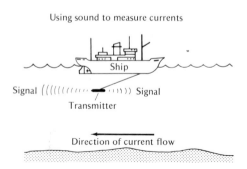

Using sound to measure currents

then tracked by the ship, and a continuous record of the current speed and direction can be made.

WATER-SAMPLING BOTTLES

One of the earliest examinations of water specimens successfully removed from the ocean depths was made in 1877 by means of the Sigsbee water bottle, a device invented by C. D. Sigsbee of the U.S. Navy. The bottle had open valves at either end. A propellerlike device turned and sealed the bottle at the proper depth as the bottle was returned to the surface.

Today, oceanographers use an improved water sampler called the *Nansen bottle* (Figs. 2-7 and 2-8). These bottles are attached to a wire at a variety of depths. Valves at either end are left in an open posi-

---

2-7 A Nansen bottle being tripped. *(After the USNOO)*

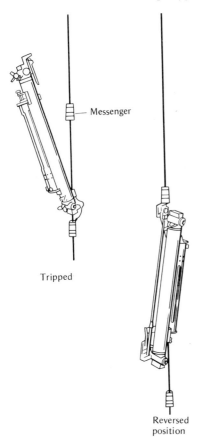

Messenger

Tripped

Reversed
position

2-8 A Nansen bottle being recovered. *(NOAA)*

tion until the wire and its bottles are set in place. A weighted *messenger* is dropped along the line to trip over the first bottle on the wire. As the bottle trips over, the valves at either end close off the contents and trap the water within. This bottle sets off a second messenger, which drops down the line to the next bottle. As each bottle is tripped in turn, a messenger is dropped to the next bottle on the line. Thus, a large number of samples can be collected on a single line. Each sample represents the water characteristics from a different level.

OTHER SAMPLING TECHNIQUES

There are a number of devices by means of which samples of water from a variety of depths can be examined. One type of water bottle consists of a tube with rubber stoppers at either end. This operates in a fashion similar to the Nansen bottle. In shallow waters, samples are taken by a pump which draws water through a hose. Electronic sensors in the device automatically record the temperature, acid-base balance of the water (pH), gas concentrations, and other characteristics.

THE MEASUREMENT OF SPECIAL CHARACTERISTICS OF SEAWATER

Electronic and visual observations are made for several other physical characteristics found in the sea. Among these are the *turbulence* and *transparency* (in respect to light) of the water. Turbulence is a difficult determination to make, as records of the temperature, salinity, and velocity factors must be made. Each factor is affected by the turbulence. The factors change continuously as they are influenced by the amount of turbulence of the water. The variations in the fluctuations of the water are recorded electronically, and a continuous record is made for each factor.

The degree of transparency of the water is measured with a *Secchi disk,* a small black and white disk about 8 to 18 inches in diameter (Fig. 2-9). The disk is lowered into the water, and a notation is made of the depth at which it disappears from view. The depth of disappearance is a determination of the transparency of the water. In clear

2-9   A Secchi disk.

water the disk can be seen to depths over 50 meters; in coastal waters the disk disappears at about 1 meter. The average depth along most coastlines is about 10 meters.

A light source and photocell are combined in a device known as a *hydrophotometer* for transparency determinations. As the suspended material in the water cuts off the light from the photocell, a reading is automatically made of the depth and the degree of transparency is calculated.

TIDES AND WAVES

In addition to currents, tides and waves (discussed more fully in Chap. 8) are periodic and repetitive motions made by the water of the ocean. These, too, are measured for a variety of research purposes.

Tides

Tides move rather slowly and, as a result, an accurate profile of the daily fluctuations in water level is easily made. Tide heights are measured as a continuous movement by means of *tidal staffs,* which are placed into the water in a vertical position. A float moves upward along the staff and a continuous and permanent record is made by means of an electric connection between the float and a recording device.

Current meters also are used in tidal channels. In these areas, the flood-tide and ebb-tide velocities are determined in exactly the same manner as the currents on the open sea.

Waves

Waves are, of course, more pronounced and periodic than the tides in many parts of the world. In these, the factors with which the oceanographer is concerned are the height of the waves, their period, and length (see Chap. 9). These are usually determined so that a profile for a large unit of time can be obtained (see Fig. 2-10).

Wave staffs are used in measuring wave height. These vertical rods generally use an electric signal to transmit the changes in each wave height. Marked rods, another form of wave-measuring device, are read by an observer to obtain a close reading of the wave heights over a specified period of time.

As waves move into shallow water, they create changes in water pressure on the bottom. Pressure gauges placed on the bottom in these shallow regions give variations which are the direct result of changes in the wave characteristics of height and length. Thus, the characteristics can be conveniently obtained, especially in coastal regions.

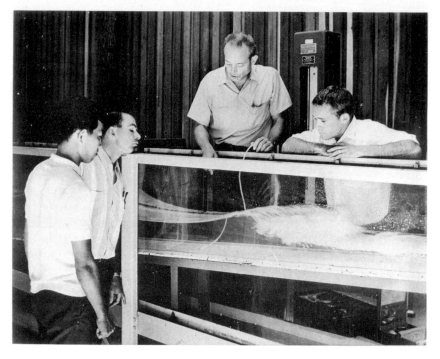

2-10   A wave tank; used to examine the behavior of wave forms.

BIOLOGICAL AND GEOLOGICAL DATA COLLECTING

In addition to the concerns of the physical oceanographer, marine biologists and geologists also require data and samples for their specialized investigations. Examinations involve a variety of methods for the return of the specimens to the ship. Depth measurements reveal the contour of the ocean floor, required for a true picture of the nature and formation of the geological features among which marine organisms live. A great number of devices are used to collect these data. Many techniques have been devised to measure the depth and contours of the ocean floor.

BIOLOGICAL SAMPLING
Biological samples have been collected from the ocean floor for many decades. From 1880 to 1930, the expeditions of the English, Norwegian, German, and other investigators collected thousands of biological specimens for laboratory examinations. Most specimens

were collected from the Atlantic Ocean, but collections also were obtained from the Pacific and Indian Oceans. It was during this same period of time that many nations founded marine stations for oceanographic research.

One of the most common methods of biological sampling is to use nets. Nets and hooks have been used since prehistoric times, and modern techniques are variations of one of the oldest means for capturing fish and other marine life (Fig. 2-11).

Nets, of course, come in all sizes and shapes. One of the most useful is the *plankton net,* generally a cone-shaped net made of extremely fine mesh (Fig. 2-12). The net collects the tiny organisms that make up the vast array of planktonic life-forms in the ocean. This examination is taking on greater importance as man searches for a larger source of food to feed the world's growing population.

A small propellerlike metered device is often attached to the front of the plankton net as it is dragged through the water. The water passing through the net turns the propeller, and an accurate measurement of the amount of water moving through the net is made. This gives the marine biologist a measure of the quantity of plankton in

2-11   Fishing methods have remained largely unchanged throughout the ages. *(FAO-UN)*

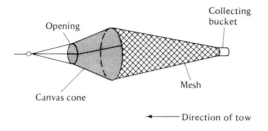

2-12   A plankton net.

---

the unit volume of water, thus assuring a better measure of the concentration and dispersal of these organisms.

The closing net is one of the earliest devices developed for the collection of deep-water organisms. This fine-meshed net is hauled behind the ship. The net is closed when it is first dropped into the water, but when it reaches a predetermined depth, a weight dropped along the line opens the net. Nets of this type have been used since 1884 at depths ranging down to 4,000 meters (2½ miles).

Trawls, which are used to capture many bottom specimens, resemble dredges which drag along the bottom as they are towed by a ship. They pick up whatever is in their path on the ocean floor.

GEOLOGICAL SAMPLING
The recovery and examination of the materials which lie on the ocean floor are equally important to geologists and biologists. Several devices, simple and complex, have been developed for this purpose and are in common use throughout the world today.

The bottom samplers used by marine geologists consist of two basic types. One, the grab samplers and dredges, retrieve samples of the upper surface layers. The other type, corers and drills, recover samples from greater depths in the sediments. The cores represent samples exactly as they were laid down by the original processes.

Grab Samplers and Dredges
In order to obtain a rough analysis of the ocean floor, *grab samplers* (Fig. 2-13) are used to recover geological samples. They have a pair of jaws similar to those on a steam shovel. A smaller form, called the *orange-peel sampler,* resembles the pushed-back sections of an orange peel. The sampler bites into the sediment on the ocean floor and is then returned to the ship for later investigation of the sample. Both rocks and living organisms can be captured by such devices.

Another device, called the *dredge,* is used to recover rock samples

2-13   Grab sampler. *(NOAA)*

---

from the sea floor. It consists of a large chain mesh attached to a box-like metal opening. The front edge is rather sharp and prevents the device from snagging. The dredge utilizes a 1- to 3-foot-wide pipe with a grate at the open end to catch the larger samples which pass into the pipe.

Core Samplers

Grab samplers and similar devices are effective in collecting samples of the floor, but they present one major problem: the sample is not recovered intact, and scientists cannot study the sample exactly as it was laid down. In order to collect undisturbed samples of sediment, *core samplers* have been developed and are currently used by oceanographers to collect sediment intact from every ocean of the world (see Fig. 2-14).

In its simplest form, a corer consists of a long, hollow metal tube with a weighted upper end. When the tube is dropped overboard, the weight drives the open end into the sediment and the sediment is then returned intact and undisturbed. The tube usually has a plastic liner for easy removal and storage of the core. Fins help guide the tube on a straight path. Without the fins, the tube would not fall true, much as an arrow needs feathers to fly true.

Modern corers have a smaller release weight which allows the corer to free-fall the last few feet to the bottom; at the same time it takes a small core of the upper few inches of the surface. Within the larger

---

2-14   A 6-foot gravity corer. *(Scripps Institution of Oceanography)*

corer a piston rides upward as the tube is driven into the sediment. The piston uses water pressure to overcome friction between the corer and the sediment and helps to collect a more complete sample. The piston also prevents disturbance of the sample (see Fig. 2-15b). Sometimes, a small glass ball is utilized in the core. As the corer, called a *ball breaker,* makes contact with the bottom, the glass ball breaks and the sound, picked up by the ship's underwater microphones, is used to determine the exact depth at which the core was recovered.

In recent years, cores over 60 feet long have consistently been recovered with the use of the piston corer. Each core represents the millions of years of Earth history that were required to deposit the sediment. These cores, especially those taken in regions of the floor where the history of the area is much older, reveal a great deal about the life-forms, climate, and evolution of the sea.

DEPTH MEASUREMENT AND GEOPHYSICAL TECHNIQUES

Due to the tremendous pressures found in the sea, man has not been able to descend into the oceans to directly measure the precise depth and characteristics of the sea floor. Depth measurements, which are accurate to within a small fraction, are made by various physical and electronic means. The measurements are used to produce depth charts of the ocean floor which reveal the various contours and features along the ocean bottom.

DEPTH MEASUREMENT

The measurement and recording of *depth soundings* were first made by dropping a heavy rope and weight over the side of the ship. The rope slackened when it reached bottom, and the length of rope which had been played out represented the depth of that part of the floor. The depths were then used to draw bottom charts. Maury used data from such techniques to draw a bathymetric chart of the North Atlantic Ocean with contour lines at 1,000, 2,000, 3,000, and 4,000 fathoms. This chart prepared by Maury in 1852 is shown in Fig. 2-16.

In 1875, Rear Admiral C. E. Belknap, aboard the ship *Tuscarora,* substituted a piano wire for the rope to obtain depth measurements. This was one of the first improvements in the methods used previously.

Soundings with wire and rope proved unreliable since it was difficult to determine exactly when the line hit bottom. At times, deep currents dragged the weight sideways, and false readings were ob-

(a)

2-15   (a) A corer lowered over the side of the ship. (NOAA)

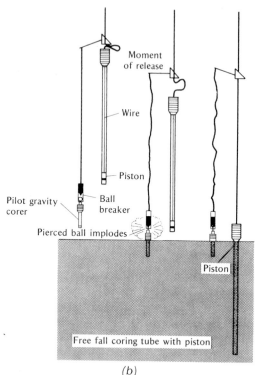

(b)

2-15 (b) Diagrammatic sketch of a corer in operation. (After Francis P. Shepard, "Submarine Geology," Harper & Row, Publishers, Incorporated, New York, 1963)

tained. Today, more reliable depth measurements are used.

Echo Sounding

Depth sounding, or *echo sounding,* is now used to determine the depths of the ocean floor. This method employs a sound pulse sent from a transmitter which is reflected from the bottom to a hydrophone, or underwater receiver, carried by the ship. The soundings are recorded on a revolving strip which gives a continuous profile of the details of the topography of the ocean floor.

The depth of the region is calculated by knowing the speed of the sound wave through the water. By utilizing the speed of the sound wave through the water and the time between the ping and the returning echo, the depth can be calculated. Different frequencies are refracted by various layers of the floor. Thus, a continuous profile can be made of the depth, and the nature of the layers of sediment can be analyzed at the same time.

Bathymetric Charts

Accurate depth measurements allow oceanographers to draw charts of the ocean floor which show the depths at each level. These *bathymetric charts* utilize a series of lines drawn through all the points on the map which are at equal depths below sea level. The lines, or *isobaths,* appear on the map as irregularly shaped lines which follow the irregular formations of the ocean-floor features. The isobaths are also called *contour lines,* as their formation is controlled by the shape or contour of the ocean floor.

Seismic Shooting

Seismic shooting (Fig. 2-17) is a sophisticated method of recording not only the depth of the ocean floor but also the thickness and type of sediments and rock layers that form the ocean floor. This technique involves the release of an explosive charge over the side of the ship.

2-16 A bathymetric chart prepared by Maury in 1852. *(USNOO)*

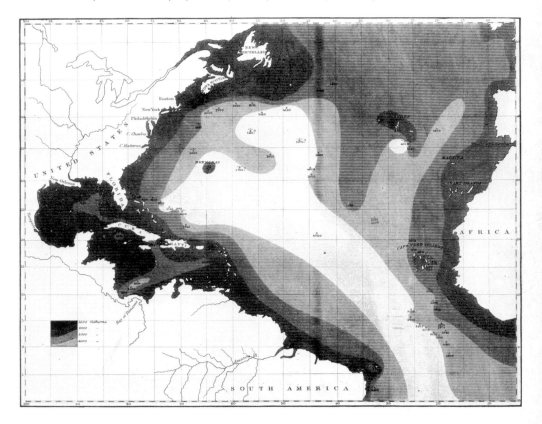

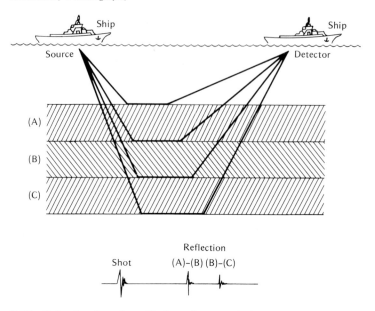

2-17 Seismic shooting. Various layers of the ocean floor refract seismic waves to give a representative cross section of the bottom.

The energy waves are reflected or bounced back from the bottom. The hydrophone receives the energy waves and transmits them to a recording device.

The energy waves that enter the layers of sediment and rock of the floor are refracted, or bent, by the various materials constituting the ocean floor. As each wave is refracted and received by the ship's devices, analyses can determine the nature of the materials of the ocean floor (see Figs. 2-18 and 2-19).

2-18 Echo sounding is used to record accurate depths.

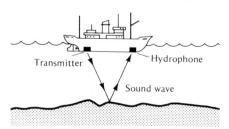

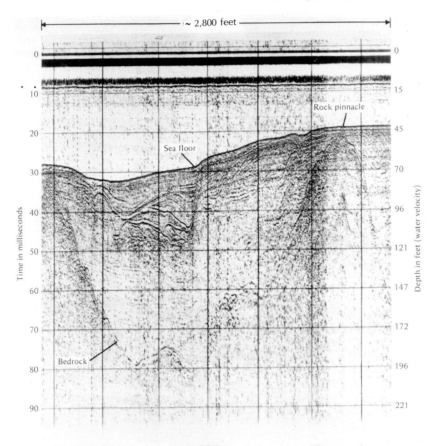

2-19 A seismic record of geologic interfaces between various sands and gravels. *(EG&G, Inc.)*

PHOTOGRAPHY

Photographic techniques currently are used as an adjunct to other sampling and measuring methods. The turbulence and transparency of water affect the success of photographic analysis of the ocean bottom. Successful photographs have been made underwater since 1893, but these early photographs were not of the highest quality. The photographs were made one at a time.

Most of the progress in underwater photography has been made since 1940, thanks to the introduction of high-speed films, automatic cameras, and light sources. Today, hundreds of photographs are made in a single descent, and many of them are in true color (Fig. 2-20).

Often, a camera will be tripped in the opening of a grab sampler just prior to the removal of a geological sample of the floor. The photograph thus shows the floor before it was disturbed by the sampler. Many bottom photographs indicate the wide variety of life on

2-20   Underwater television camera and tripod before launching over the side. *(Deepsea Ventures, Inc.)*

2-21   Tracks of living organisms taken by the R.V. *Vema* at depths of *(a)* 1,700 fathoms at 14 °6 'S, 96 °10 'W; *(b)* 1,800 fathoms at 16 °59 'S, 117 °53 'W. *(Lamont-Doherty Geological Observatory)*

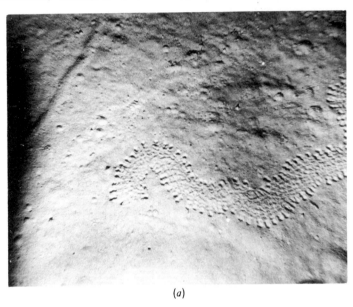

(a)

(b)

the floor in the sediment, as evidenced by the numerous animal trail marks and other formations (Fig. 2-21).

Thus, although man has not been able to descend into the depths of the ocean until recently, he has managed to develop techniques for collecting large amounts of data. His ingenuity has been the major means whereby a huge array of devices and scientific techniques have been developed to allow for indirect measurement of all the features of the oceans. Now that man is developing methods by which he can begin to visit the ocean floor, many as yet unexplored features will be revealed.

SUMMARY QUESTIONS

1. What are the primary kinds of data collecting utilized by oceanographers? What factors cause oceanographers to use this type of data gathering?

2. Describe the chief methods of collecting data in physical oceanography.

3. What advantages has the corer over the other types of geological sampling of the ocean floor? Why is core sampling one of the most significant types of data collecting in modern oceanography?

4. Describe the methods by which we profile the ocean floor and make cross-sectional analyses of the floor.

5. What major factors determine the density of water masses? How do they affect density?

6. Explain "dating" of water masses. What does dating mean?

7. What factors have prevented man from making firsthand direct observations of the ocean depths?

8. Describe the difference between simultaneous readings, continuous studies, and single-source observations. How is each accomplished?

9. Pure water is not a conductor of electricity. How does an electric current pass through a mass of water? How does this permit the measurement of a physical property of water?

10. Describe the action of a reversing thermometer and how it fixes its reading. Why is this type of device necessary?

11. Explain how three different methods may be used to measure current action. Describe the general function of one method of tracing subsurface currents.

12. How do grab samplers and dredges operate? In what way are they useful for giving the general characteristics of an area?

13. Compare and contrast the information derived from echo sounding and seismic shooting.

14. What information has been derived from advances in underwater photography? List some contributions in two different areas of investigation.

15. What information is included on bathymetric charts? How are these data listed on the chart?

16. What is the definition of density?

17. What is a wallow float? What is its purpose?

18. What does a wave staff determine?

19. What does a bathymetric chart reveal?

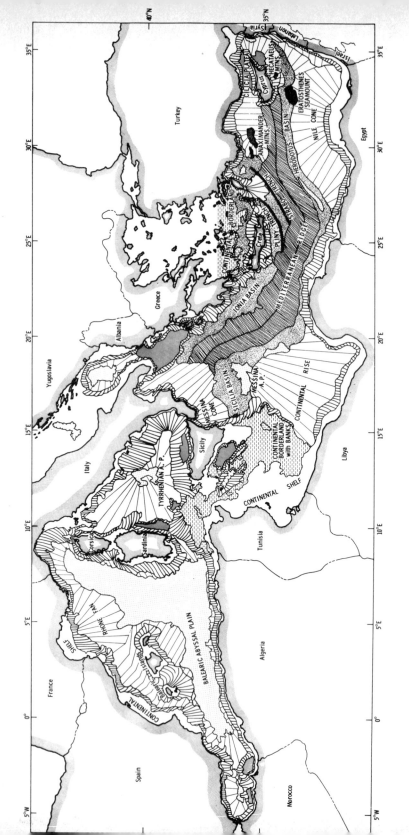

3 | OCEAN-BASIN STRUCTURE

One of the most striking discoveries of oceanographic investigation in the last generation
was the true nature of the ocean floor. Once
considered to be a flat, featureless plain, the
topographic features of the ocean floors are
now known to be more varied and more
numerous than the more familiar continental
surfaces. The nearly 71 percent of the Earth
elevations and depressions represented by the
ocean-floor relief are hidden beneath enormous
bodies of water. Thus, only in the last few
decades have the uniformity and extent of
these geological structures become apparent.

Geophysical measuring and sampling
techniques lead us to believe that the oceans
have evolved throughout the course of time.
While we know that ocean basins have always
been features of the planet Earth, the floors of
the ocean basins undergo numerous changes
and alterations through various geological
processes. As a result, major dissimilarities
exist between the ocean-basin topography and
that of the continents.

In comparison to the ocean basins, the
continents show a rather moderate relief, the
majority of the elevations and depressions
falling within a rather narrow range. However,
the ocean floors are more irregular, possessing
numerous large mountains and deep trenches.
For example, Mt. Everest, the highest elevation
on land is approximately 28,000 feet while the
deepest part of the ocean floor is the 36,000-
foot chasm known as the *Challenger Deep*.
Sea-floor features are affected by ocean
currents and the constantly shifting sediments,
while continental features are altered primarily
by atmospheric effects (see Fig. 3-2).

A physiographic diagram of the Mediterranean Sea. *(Texas Instruments, Inc.)*

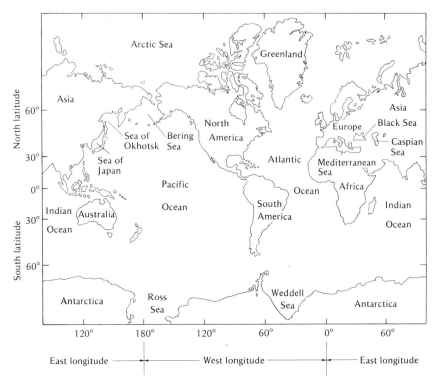

3-1   The worldwide distribution of water and landmasses.

THE EARTH'S STRUCTURE

From data collected as earthquake energy waves move through the Earth we know that the Earth is made up of several layers of rock materials of varying density. The waves, which travel both through the Earth and on the surface, have given us the limited information we possess about the internal structure of our planet. When the earthquake waves travel through the different materials making up the Earth, some of the wave energy is deflected while the remainder continues through the Earth. The record of these deflections indicates a change in the physical and chemical properties of the compositional material of the Earth. These boundaries are known as *discontinuities* (see Fig. 3-3).

Not all earthquake waves travel at the same speed, nor are they all necessarily capable of traveling through the same materials. Data regarding the patterns of different waves enable us to suggest the compositional makeup of the Earth. During periods of seismic activ-

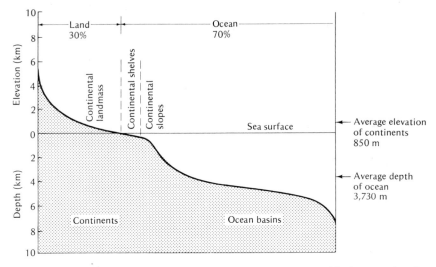

3-2 A hypsometric curve showing the elevations above and below sea level.

ity, an analysis of the wave patterns used to measure earthquake activity reveals the presence of two basic types of earthquake waves, called *body waves* and *surface waves* (see Fig. 3-4).

All body waves travel through the interior of the Earth, but some travel through only certain types of material. P waves, or longitudinal waves, can travel through any material—solids, liquids, or gases. Since these waves move faster than any other type, they are the first to arrive at a recording station. Hence, these waves are often referred to as *primary waves.* Another form of body wave is the S wave. S waves are transverse and can penetrate only the solid material of the

3-3 An Earth cross section showing the location of the Mohorovičić discontinuity.

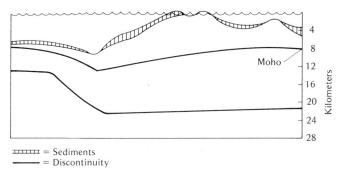

Earth. Since these waves travel at a speed about one-half that of P waves, they are the second set to be recorded on the seismograph and are called *secondary waves* (see Fig. 3-4a).

Surface waves cannot pass through the Earth but move along the surface. They are of two basic types. The *Rayleigh waves* cause particles to vibrate in a circular pattern as they pass, while *Love waves* cause particles to oscillate at right angles to the direction of wave travel. These surface waves are the last to arrive at the recording station (see Fig. 3-5) because they travel a longer distance at a slower speed.

Studies of the refraction of earthquake waves passing through the

3-4 *(a)* P waves and S waves as they appear on a seismogram.
(b) The behavior of the various seismic waves as they are refracted reveals the nature of the Earth's interior.

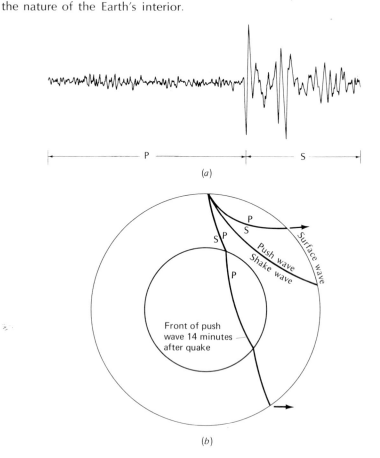

(a)

Front of push wave 14 minutes after quake

(b)

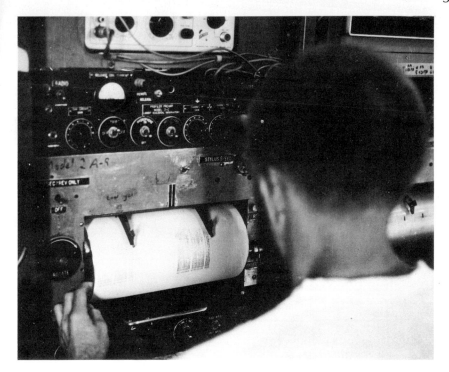

3-5 A seismic profile recorder used to determine the cross section of the ocean floor.

Earth reveal the presence of a shadow zone. This zone is a region located almost directly opposite the focus of the earthquake which receives no S waves. Since these waves do not pass through semisolid or liquid materials, their absence in the shadow zone leads scientists to conclude that a portion of the Earth's core is semisolid in nature. These waves are not transmitted through the semisolid outermost layer of the core and are blocked from the shadow zone. P waves, in turn, are refracted, or bent, on either side of a line perpendicular to the focus (see Fig. 3-4b). The refraction also prevents P waves from entering a part of the shadow zone. This unusual set of circumstances allows scientists to deduce the presence of several spheres beneath the Earth's crust, the core and the mantle layers (see Fig. 3-6).

The deepest layer at the center of the Earth is the solid core. Surrounding the *inner core* is another shell of semisolid material called the *outer core*. Although the composition of the inner core located 5,000 kilometers (3,160 miles) below the Earth's surface is not ex-

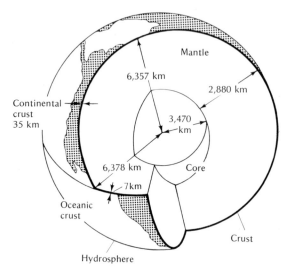

3-6   The Earth in cross section. *(After E. C. Robertson, "The Interior of the Earth: An Elementary Description," U.S.G.S. Circular 532, 1966)*

actly known, it is thought to have a nickel-iron content. This would account for the high density of 11 to 12 times that of water attributed to the core. Above the core we find a second major layer, the mantle. The mantle, which extends to a depth of approximately 2,900 kilometers (1,800 miles), is a solid having a density lower than that of the core. The outer shell covering the Earth is called the *crust*. It is separated from the mantle by a worldwide discontinuity known as the *Moho*, named for the Yugoslavian geophysicist Mohorovičić, who first demonstrated the existence of the boundary in 1909.

ROCKS IN THE EARTH'S CRUST

The rocks of the Earth's crust fall into two general categories. The typical *continental crust* is composed of a granitic rock underlain by a second layer composed primarily of basaltic rock. The granitic rocks do not occur in the oceanic crust, where one finds a single layer composed of basalt. The major component of the upper continental crust is rich in silicon and aluminum and is called *sial* (see Fig. 3-7). The basaltic layer beneath the sial and in the ocean basins is rich in magnesium and iron and is called *sima* (see Fig. 3-8). It is difficult to get precise data on the total thickness of the crust, but it is certain that the crust of the Earth is thickest beneath the continents. From a study of seismic waves the continental layers are

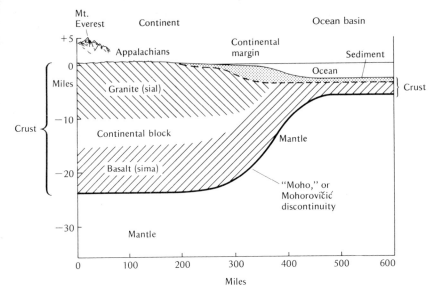

3-7 The continental crust.

thought to be approximately 33 kilometers (20 miles) thick, being 20 kilometers (12 miles) deep at the shallowest point and 64 kilometers (40 miles) at the deepest boundary. The single-layered crust beneath the oceans is considerably thinner; measurements reveal it to be only 5 kilometers (3 miles) thick beneath some portions of the Pacific Ocean. Geologists maintain that the crust of the Earth floats on the mantle with the "lighter" granitic rocks floating higher than the "heavier" basaltic crust of the ocean (see Fig. 3-9).

The granitic rock materials are not exposed at all places on the surface of the Earth because of the covering of soil and other rock types deposited since the formation of the continents. The same holds true for the basaltic materials of the ocean basins, which have been cov-

3-8 The oceanic crust.

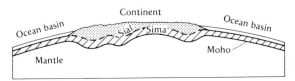

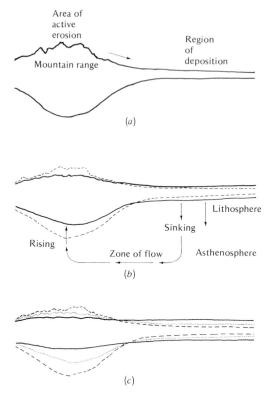

3-9   Isostatic adjustment in crustal rocks.

ered by hundreds of feet of sediment in many places. In some places, outpourings of volcanic basalt have been built up from the floor; for example, the Hawaiian Islands rise 11,000 meters (35,000 feet) above the surrounding ocean floor.

ISOSTASY

Like all geological structures, the rocks of the continents and islands are continuously subjected to weathering. These processes wear down the elevated landmasses, and eventually the eroded material is carried out to sea. This continuous weathering and concurrent erosion tends to produce changes in the mass of material deposited on various parts of the Earth's crust. However, changes in the mass of the Earth's crust are accompanied by movements which tend to equalize the change in mass. This tendency toward an adjustment in equilibrium is known as *isostasy*. Isostasy, in conjunction with other geological forces, such as volcanic action and earthquakes, produces a number of irregularities in the crust of the Earth.

The balance known as isostasy may be upset by the removal of material from one area of the Earth's surface and its deposition in another region. For example, the changes in sea level that occurred during the glacial stages of the Earth's history caused isostatic adjustments in the Earth's crust. The water that evaporated from the oceans was returned to the land in the form of rain, snow, and other forms of precipitation. The formation of ice sheets caused an overloading of some areas, which in turn caused the material beneath the crust to begin to flow toward underloaded areas. The underloaded areas began to rise, and the process, which is still under way, will continue until balance is restored. Similarly, when a large mountain range is eroded and the eroded material is deposited in adjacent regions, an isostatic adjustment will occur.

Recent research has attempted to discern the precise effects of isostatic adjustment. Careful measurements of the Scandinavian landmass reveal that it is slowly rising, in parts, at a rate of 1 to 3 feet per century. This region, once depressed by the glaciers of the last Ice Age, is now rising due to the melting of the ice and is responding to the reduction in weight. The rocks move slowly, however, and the adjustment is not yet complete.

The term isostasy, which comes from the Greek meaning "equal standing," is not a force but a state of equilibrium maintaining a balance between materials of greater and lesser densities. Mountains of lighter material, floating on denser underlying material, cause the base of the crust to be depressed. The light mountain mass, usually granite, extends far down into the Earth to form roots of lighter material. Beneath the oceans, the denser basaltic materials are very shallow. The amount of "light" material therefore grades downward, being deepest beneath the mountains, shallow beneath plains and valleys, and almost nonexistent beneath the oceans. This phenomenon accounts for the granitic crust of the continents atop the basaltic crust. Further, the base of the crust is found at greater depths beneath the continents than beneath the ocean basins.

OCEAN-BASIN TOPOGRAPHY

The topography and structure of the oceans may be divided into three major units: the continental margins, the deep ocean-basin floor, and the mid-ocean ridges. In turn, these units may be further divided into subunits, including the continental shelves and the continental slopes, which constitute the margins of the continents; the continental rises, submarine canyons, deep-sea trenches, abyssal plains, mid-ocean canyons, abyssal hills, and fracture zones, which constitute the ocean-basin floor; and finally, the worldwide ridge system and rift valley. Although our knowledge of the topography of the ocean floor is far from complete, modern researchers are developing a recent and accurate physiography of the oceans.

THE CONTINENTAL MARGINS

The *continental shelf* is a part of the transition zone, the continental margin, which exists between the continents and the ocean basins (see Fig. 3-10). The continental shelves fringe the continents and slope gently toward the ocean basins. The width of the shelf varies from a few miles to over several hundred miles, especially in regions which once were glaciated. Bordering nearly all landmasses, the shelf represents a submerged extension of the continents and is considered part of them. The shelves are relatively shallow, for the most part less than 600 feet deep. The rocky shelf, which has been modified in part by marine erosion, is covered by a variety of sediment types and underlain by a continental type of granitic crust. The *shelf break* marks the seaward extent of the continental shelf, sharply dividing it from the continental slope.

The *continental slope* (marked by a steep gradient of 1:40 compared to 1:1,000 of the shelf) extends to a depth of 1 to 2 miles. Bordered in places (especially in the Pacific basin) by the deep-sea trenches and fault surfaces, the slope represents one of the most marked relief features of the Earth. The slope is scarred by submarine canyons which resemble gullies and sometimes rival the Grand Canyon in size. These canyons, probably cut by currents of muddy water, are V-shaped valleys in the mud and rock outcrops of the

---

3-10   The general configuration of the continental margins.

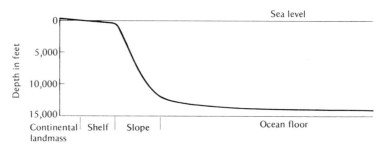

slope. The mud moves downslope as a cohesive unit with a density greater than that of seawater. The movement of this mud slurry, called a *turbidity current* (see Chap. 8), is activated by a sliding or slumping of sediment along the continental margin. The sediments are set in motion when physical forces such as the tremor of an earthquake, the high transport rate of sediments by the continental rivers, or any other factors affect their stability.

Turbidity currents, which are the main means of transporting sediments to the abyssal plains, have produced a number of consequences of direct concern to man. For example, Bruce Heezen and Maurice Ewing, scientists at *Lamont-Doherty Geophysical Laboratory* of Columbia University, pointed out that the Grand Banks earthquake of 1929, which resulted in the downslope movement of sediment, produced a turbidity current which consecutively broke the trans-Atlantic cables in that region. They theorized that the slumping produced by the earthquake caused an increase in density when the sediments mixed with the surrounding waters. The thick slurry moved downslope, continuing to accelerate as it moved, until the slope of the sea floor decreased, resulting in a slowing of the current and the eventual settling out of the suspended sediments. From a record of the time sequence of breaks in the cable, it was possible to determine the speed of the turbidity current. Investigation reported speeds as high as 2,000 centimeters per second, confirming that the currents can reach extremely high velocities. To confirm the theory that the Grand Banks cable was destroyed by the turbidity current, core samples were taken. The cores revealed that shallow-water sediments had been transported from the edge of the shelf to the deep water by the turbidity current. The Grand Banks event is not an isolated example. Cable breaks elsewhere may be attributed to the rapid movement of turbidity currents.

The sediments from the shelf are carried into the deep ocean, carving a path of deep canyons and altering the slope topography in various ways. At the foot of the continental slope sediment has accumulated to form the continental rise. However, the origin of submarine canyons is still not completely understood. Other researchers have suggested that these canyons are drowned river channels which formed during periods of reduced sea level when the present continental shelves were exposed.

The *continental rise*, although it does not exist in all parts of the world, generally blends gently into the abyssal plains of the deep ocean. This gradual slope, from 1:1,000 to 1:700, is cut by deep canyons which serve as channels for the seaward transport of sediment.

The smooth surface of the continental rise varies in width from a few miles to a few hundred miles. This apron of sediment may reach depths as great as 17,000 feet at its outer edge.

The type of continental margin varies with geographical location. For example, the margin off the eastern coast of the United States is rather wide, while off the western coast it is extremely narrow. On the eastern coast of Newfoundland the shelf averages more than 200 miles in width. While this is the widest point along the Atlantic coast, other areas are also quite wide. The eastern continental shelf of the Atlantic Ocean narrows slightly, ranging approximately 100 miles wide north of Cape Hatteras to only 19 miles off Cape Hatteras. It immediately widens to 60 miles off the Georgia coast and narrows once again, to disappear south of Palm Beach, Florida. On the Pacific coast of the United States the shelf (when it exists) averages between 10 and 20 miles in width. The shelf off San Diego is only 10 miles wide. Farther north, the shelf narrows to only a few miles near Long Beach, where it begins to widen, reaching a maximum of 15 miles off San Francisco. The shelf once again narrows to about 10 miles off the coast of northern California and widens to about 20 miles off the Oregon and Washington coastlines. A wide margin generally contains all three zones—the shelf, slope, and rise—while narrow margins, typical of areas of recent seismic activity, often descend more sharply directly into the ocean floor. These narrow regions often are marked by deep trenches, some as deep as 7 miles.

THE OCEAN-BASIN FLOOR

The ocean-basin floor extends seaward from the margins to the mid-ocean ridges and comprises three categories of topographic features: abyssal plains, oceanic rises, and seamounts (see Fig. 3-11).

Abyssal Plains

The abyssal plains situated at the base of the continental rise are flat, almost featureless, plains having a shallow gradient of 1:1,000. Sediment from the continental rise carried down by turbidity currents has buried any preexisting rough topography. Abyssal plains represent the flattest part of the Earth's surface and are found in all oceans. However, they are most common in the Atlantic and Indian Oceans and are quite rare in the Pacific Ocean.

Associated with abyssal plains are *abyssal hills*. Some of these small oval-shaped hills were formed from masses of continental sediment which buried the topography. Others, located in the central portion of the ocean basins, are sediment-covered igneous-rock masses.

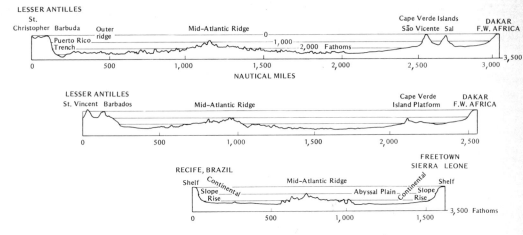

3-11 Deep-sea floor topography. *(From B. C. Heezen, GSA Special Paper 65, 1959)*

Oceanic Rises

Another characteristic topographic feature of the ocean-basin floor, the *oceanic rise,* represents isolated areas reaching several hundred feet above the surrounding abyssal plain. Having a varied relief, the rise may consist of low hills to large mountains as much as 5,000 feet high. The best example of this type of ocean topography is the Bermuda Rise. Located in the North Atlantic Ocean basin, this massive rise serves as a pedestal of volcanic material upon which the accumulation of coral growth has formed the Bermuda Islands.

Trenches

In addition to the various elevations found on the ocean floor, great depressions also are to be found. These depressions, called *trenches,* usually border the continents. While they are found in every major ocean in the world, the largest number encircle the Pacific coastline associated with the area known as the *ring of fire.* The term is derived from the frequent earthquake and volcanic activity found in the vicinity of the circum-Pacific basin. The Mariana Trench, in the western North Pacific, is the deepest in the world—over 36,000 feet. Trenches are often shaped like the arc of a circle and are usually associated with an island arc on the continental side. The island arcs are long, narrow, submarine ridges rising above the level of the sea.

Seamounts

Fascinating features of the ocean basins are the seamounts, isolated peaks that rise thousands of feet above the ocean floor. Seamounts usually are steep-sided, but erosion by waves has planed off the sharp peaks in many places. This leaves a flat-topped, sunken seamount, called a *guyot* (see Fig. 3-12), named after the nineteenth-century Swiss-American geologist Arnold Guyot. It is thought that seamounts and guyots are volcanic cones built up from the ocean floor by a series of volcanic eruptions. Furthermore, the shape of guyots is evidence that these ancient seamounts extended above the surface of the ocean as isolated islands during periods of lower sea level. These sunken islands appear to confirm that the level of the sea has changed throughout time. Guyots may also be evidence of isostatic adjustments of the sea floor. That is, these once exposed islands may have slowly subsided, in addition to being drowned as the level of the sea rose.

THE MID-OCEAN RIDGE SYSTEM

The most remarkable physiographic feature of the ocean floor is the Mid-Ocean Ridge System. Extending throughout the world ocean, this continuous mountain range circles the Earth for 40,000 miles or more. Reaching into every ocean basin, the ridges are between 600 and 2,500 miles wide with a relief of 6,500 to 13,000 feet above the ocean floor (see Fig. 3-13). Portions of the ridge system may protrude above the sea as island peaks. Composed of a combination of volcanoes and fracture zones, the ridge system represents an area of great seismic activity. Much of the topography of the ridge has been

3-12 A guyot.

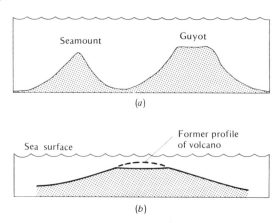

(a)

(b)

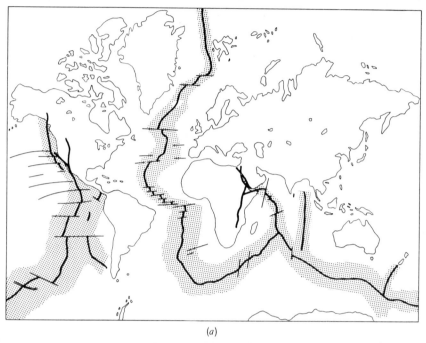

(a)

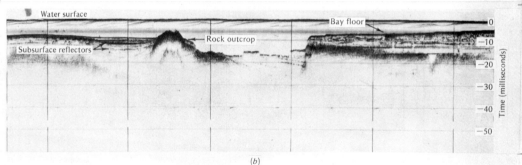

(b)

3-13 (a) The worldwide ridge-rift system.
(b) A profile graph. (U.S. Naval Oceanographic Office)

formed by the abundant earthquake activity. In addition, evidence of
the higher than normal heat flow indicates the influx of the hot,
molten material, or the *magma*, from beneath the crust. For example,
the fracture zones represent an area in which the oceanic crust splits,
resulting in movement of rocks on either side of the fault. Many
faults of this type are perpendicular to the major ridge system. The
Mid-Ocean Ridge System has a characteristic depression, called a

rift valley, running down the middle of the ridge crest. Many other features, such as a cliff face resulting from movement of the undersea crust, parallel the rift. These fault *scarps* suggest that faulting of the crust was responsible for the formation of the ridge. The discovery of the ridge has also revived the hypothesis of drifting continents, that is, that the present continents once were joined together, subsequently broke up, and slowly moved to their present positions (see Fig. 3-14).

SUMMARY QUESTIONS

1. In what way do the ocean basins and the continental masses differ? Are there any similarities?

2. Briefly describe the internal structure of our planet.

3. What is isostasy? What causes isostatic adjustments to take place on the Earth's surface?

3-14 A photograph of sea-floor sediments taken by R.V. *Vema* at 44°20′S, 54°15′W. *(Lamont-Doherty Geological Observatory)*

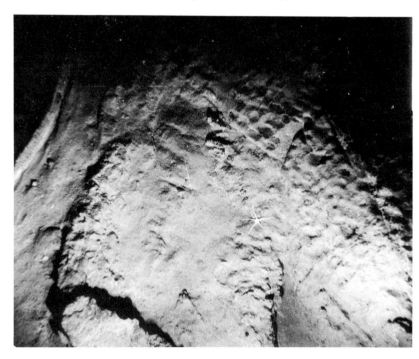

4. What features make up the continental masses?

5. In general terms, explain the variety of features (elevations and depressions) found on the ocean floor.

6. What is the nature of the ridge-rift system? In what important way may it be deemed the most significant feature of the ocean floor?

7. How are earthquake waves affected by different sections of the Earth? How does this yield data about the nature of the Earth?

8. Explain the nature of P waves and S waves and how they are indicative of the precise nature of the Earth's core.

9. Relate the differences in the Earth's crust between the ocean basins and continental masses.

10. What are the continental margins? What structures make up these features, and what are the characteristics of each structure?

11. Explain how a turbidity current may arise. How may major ocean-basin features be the result of these phenomena?

12. What are guyots? What major events in the Earth's past might they indicate?

13. What are discontinuities and what do they indicate about the nature of the Earth?

14. What is the origin of the seamounts on the ocean floor? What does this seem to indicate about the nature of the ocean basins?

15. What features are almost always associated with trenches? Describe the usual configuration of the two different structures.

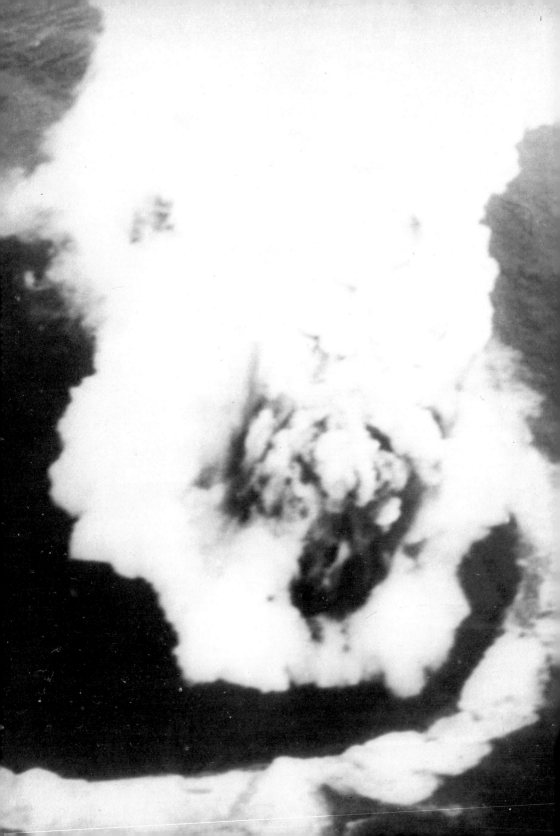

4 | THE ORIGIN OF THE OCEAN BASINS

Geologists generally agree that the water of the oceans resulted from volcanic activity, which gradually released the water trapped within the rocks of the Earth's interior. Data from radioactive dating reveal that primeval ocean basins may have formed over 3 billion years ago. Some of the oldest ocean rocks collected are pebble-bearing sedimentary rocks. Sedimentation suggests water formation, which further confirms that the oceans are at least 3 billion years old. The age of the Earth has been estimated at 5 billion years.

While the ocean basins are considered one of the Earth's oldest topographical features, it is thought that the configuration of the ocean basins has changed constantly, due to the shifting of the continental blocks. There is much evidence to show that the continents now occupy positions different from those they occupied earlier. It appears that entire sections of the Earth's crust—including continents and ocean basins—undergo a process of continuous movement. Other changes, such as those in coastlines, occur when landmasses rise and fall or when the sea surface changes in relation to a continent.

The changes in the geography of the Earth have become obvious as a result of worldwide studies of oceanic rock. The controversial theory of continental drift suggests that changes have taken place in the location of the continental blocks accompanied by alterations of the ocean floor. Attempts to correlate the rocks and fossils of different continents also seem to support this hypothesis. This phenomenon was first suggested by early

Volcano Surtsey. *(U.S. Naval Oceanographic Office)*

scientists who noted a remarkable similarity in the shapes of the continental coastlines on either side of the Atlantic. Further studies of the Atlantic coastlines, along with the parallelism of the Mid-Atlantic Ridge to the continental margins, has resulted in several arguments favoring continental drift and sea-floor spreading. This geological controversy is the subject of this chapter (see Fig. 4-1).

CONTINENTAL DRIFT

Although the term continental drift may be traced as far back as the early 1800s, the first significant data were set forth by Alfred Wegener (1880–1930), a German meteorologist, in "The Origin of Continents and Oceans," published in 1912. In an attempt to explain the distribution of Ice Age fossils and deposits, Wegener assumed that all continents had once been grouped together in a single supercontinent called *Pangaea* (see Fig. 4-2). According to his theory, Pangaea was centered around the South Pole, which was located in the vicinity of present-day Africa. Wegener further subdivided the landmass into Laurasia, in the Northern Hemisphere, and Gondwanaland, in the Southern Hemisphere. In terms of present landforms Laurasia consisted of North America and Eurasia, while Gondwanaland comprised South America, Africa, Antarctica, Australia, and India.

Wegener suggested that approximately 200 million years ago the forces associated with the rotation of the Earth caused this large supercontinent to break apart, establishing the Atlantic and Indian Oceans (Fig. 4-3). Although physicists and geologists alike strongly disagreed with Wegener's proposal, they were at a loss to explain the validity of the evidence suggesting that the continents fit together like pieces of a giant jigsaw puzzle.

As early as the nineteenth century, geologists studying the Gondwana province in east central India located a sedimentary-rock section which provided an excellent standard for comparison to other parts of the world. The plant life peculiar to this section was called the glossopteris flora, after the big-leafed fern of the same name, which was the most abundant genus present. At the same time, similar sequences of rock layers were being discovered in Australia and South Africa. Later, discoveries in South America and Antarctica revealed the same geological sequences.

When all the data from these continents are considered, they support the concept of an original supercontinent. Although Wegener's theory was growing in popularity, geologists were still looking for answers to the mounting evidence that an Ice Age had spread a

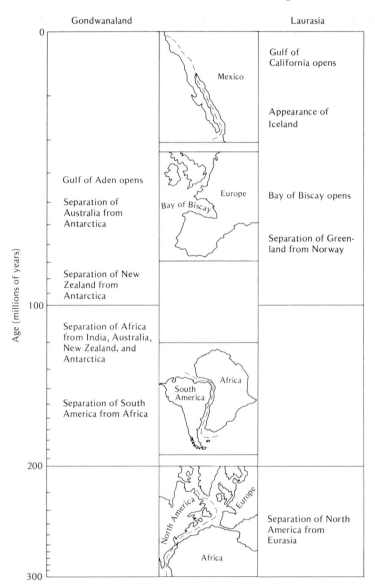

4-1 A generalized time sequence of continental drift.

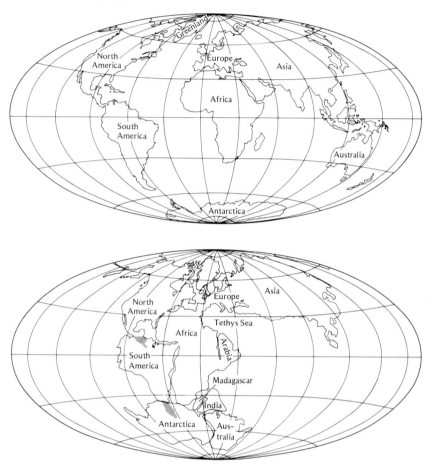

4-2 The original supercontinent Pangaea about 150 million years ago.

glacier across continents of the Southern Hemisphere 200 million years ago. According to the geologic record being established, at the same time that a glacier was moving across the southern continent, huge coal deposits were being formed in tropical forests of the Northern Hemisphere. To resolve this growing paradox a different arrangement of the continents was suggested. Alex L. DuToit, a South African geologist, wrote a book in 1937 entitled "Our Wandering Continents," in which he viewed the southern continents together at the South Pole and the northern coal forests situated toward the equator. He envisioned the eventual breakup of the southern continents with the resulting subcontinents drifting northward.

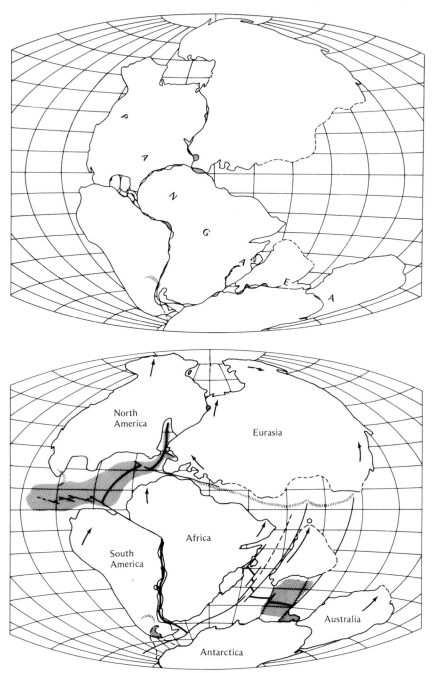

4-3 The breakup of Pangaea. *(Based on Dietz and Holden)*

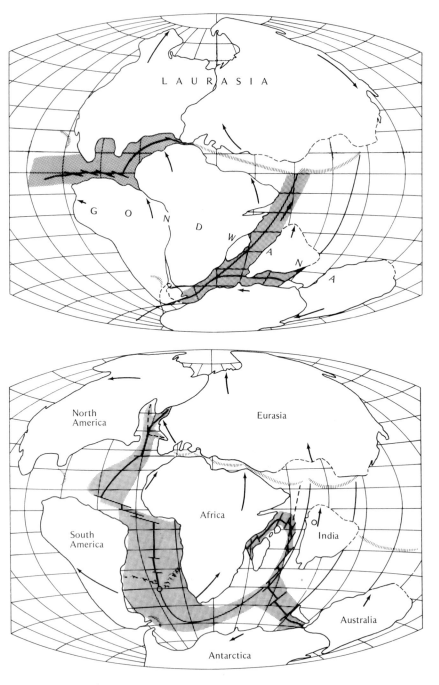

4-3 *(Continued)*

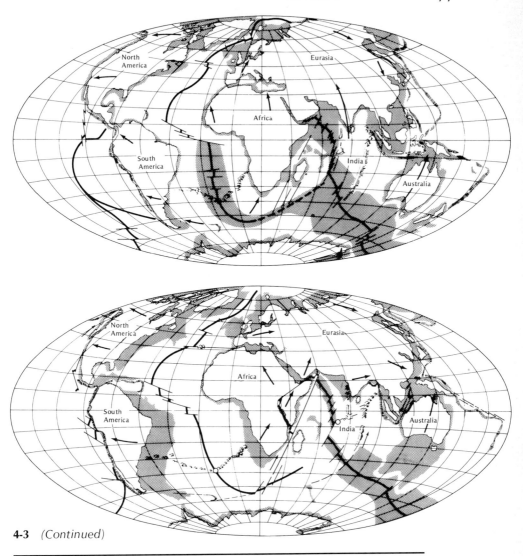

4-3 *(Continued)*

Since the glossopteris-bearing beds overlie ancient glacial deposits, several questions become obvious. The grooves made by the glacier *(striations)* indicate that the ice moved from areas that are now covered by oceans and at one time had a tropical climate. How could extensive glaciers form in tropical regions? The glacial deposits covered by beds containing the glossopteris flora also indicate the probability of continental drift. They contain rocks not found on the continent and, in some cases, rocks which match those found in outcrops on adjacent continents. How did the glossopteris flora spread across the oceans now separating these areas?

A MODERN THEORY

The attempt to answer these questions—and others like them—has led to the present theory of continental drift. In order for this drift to have occurred, the continents of the Southern Hemisphere would have had to be connected as a single landmass. Furthermore, the present geography can be explained only by the breakup and drift of the sections of the landmass.

As convincing as some of these arguments may be, the most compelling evidence that the continents were once united and are now spreading apart comes from a study of *magnetic anomalies.* These anomalies were recognized as a result of recent studies of the ocean floor.

From a study of the past position of the Earth's magnetic pole it has become evident that the pole locations with respect to the continents have migrated. The geologic time lines established for each continent are similar, even though they may show minor differences. Since Precambrian time, over 600 million years ago, the poles apparently have migrated. Further, the poles periodically have reversed. When reversal last occurred, the present north magnetic pole was located in Antarctica.

Based on the study of magnetite in igneous rocks and hematite in sedimentary rocks in many places and of different geological periods, a chart of past magnetic polarization has been established. At the time of their formation, the iron particles in lava outpourings became weakly magnetized and aligned themselves like magnets to coincide with the magnetic poles of the Earth's field. Upon cooling, the iron particles became fixed in their position. When the magnetic field of the Earth later shifted to a new position, the original alignment of the iron particles was maintained in the rocks. Subsequently, iron-bearing rock material forming after this shift in poles has particles aligned with the new poles. If the continents remained fixed throughout geological history, all the magnetic poles would line up, pointed in the direction of the present magnetic field. However, since the continents were drifting, the various rock strata indicate the relative position of the continents during the span of time (see Fig. 4-4).

At first, geologists attempted to explain this apparent changing position of magnetic fields by suggesting that the poles of the Earth had shifted many times from their present positions. It is true that

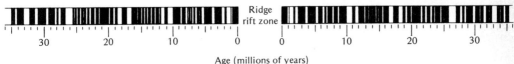

Age (millions of years)

4-4 A chart of magnetic anomalies compared from various ocean rocks.

some evidence does exist for a drifting-pole theory; however, when the continents are investigated individually, it appears that the poles took a different path relative to the different continents. This phenomenon is due to the sequence of the moving continents. The reversals of the magnetic poles are also recorded in the rocks of the ocean floor. There is ample evidence to indicate that the present positions of the continents are far from their original ones. The only phenomenon yet to be explained is the rate at which the continents moved away from each other. However, an extrapolation of the time scale established by a study of the magnetic data and supported by rock drill cores gathered during the recent JOIDES (Joint Oceanographic Institutions for Deep Earth Sampling) deep-sea drilling program has tended to establish and confirm rates of spreading (see Fig. 4-5).

Many scientists have attempted to prove or disprove the theory of continental drift. Probably one of the most conclusive pieces of evidence in support of the continental-drift theory is the mathematical construction presented by Sir Edward Bullard, professor of Geophysics at the University of Cambridge, and his associates. Bullard presented a fit of the continents using a computer to solve the problem. The fit was not made at the coastline but along an extension of the continental landmass currently submerged several thousand feet beneath the sea surface. The results were amazing. The average error over the entire construction was no more than 1° out of alignment for every boundary, when correlated with the true edge of the continent.

Incorporating much of the data collected during previous investigations, Robert S. Dietz presented a new hypothesis in 1961. This hypothesis, now generally referred to as *sea-floor spreading,* has succeeded in overcoming many of the objections to the theory of continental drift. Sea-floor spreading appears to be the most reasonable mechanism suggested to date to explain continental drift and perhaps explain why the continents began to break up in the first place (see Fig. 4-6).

4-5   A chart of Joint Oceanographic Institutions for Deep Earth Sampling (JOIDES) data.

Leg number	Area covered	General data
1	Gulf of Mexico, area east of the Bahama platform	Indicated widespread chert layers which pointed out the need for hole reentry systems. Showed existence of cap-rock formation beneath deep ocean sediments. Proved that gassy, expanding cores can generate in a deep ocean en-environment.
2	North Atlantic	Sampled the oldest available sediments in the deep sea. Tested the hypothesis of continental drift and sea-floor spreading. Sampled a complete sequence of sediments to serve as a standard biostratigraphic reference for the North Atlantic.
3	South Atlantic Ocean	Investigated the tectonic development of the Mid-Atlantic Ridge and history of sedimentation in the South Atlantic. Preliminary plots of the age of the sediments versus the distance from the ridge indicate a sea-floor spreading of about 2 cm/yr over the last 70 million years.
4	Western South Atlantic and Caribbean	Made to enhance knowledge of basins adjacent to continental margins, sea-floor spreading, nature of transverse fractures of Mid-Atlantic Ridge, and geology of island arcs and trenches, and to explore geologic history of Caribbean.
5	Northeast Pacific	Examined magnetic anomaly pattern in the Pioneer-Mendocino Fracture Zone area. Determined history of the equatorial current system and the North Pacific gyre from the sediment record.
6	Northwestern Pacific and Philippine Sea	Established the origin and history of the oceanic crust. Gathered paleontological-biostratigraphic data. Contributed to the general understanding of ocean sedimentation.

7	Western Equatorial Pacific	Studied age of crust in various parts of this region. Examined the nature and the age of the many acoustic reflectors that were observed on seismic reflection profiles.
8	East Central Equatorial Pacific	Studied geologic history of region based on sediment cores collected across the equator along the 140th meridian. Conducted an extensive investigation of the equatorial current system.
9	Equatorial Pacific: East–West Track	Pacific plates move due west (8 cm/yr to 13.3 cm/yr) from the Eocene until middle Miocene, then north at one-half former rate.
10	Southern Gulf of Mexico	Basin formed since late Cretaceous; recovered Cenozoic and late Cretaceous pelagic sediments.
11	Western North Atlantic	Determined the processes involved in forming the modern continental rise and outer ridge. Dated age of sea floor in Western Atlantic.
12	North Atlantic	Carried on routine geophysical survey and did detailed sedimentary successions for four geographical regions in North Atlantic: Labrador Sea, Reykjanes Ridge, Rockall Arc, and Bay of Biscay.
13	Mediterranean Basin	Collected many different rock types. Evidence exists for compressional tectonics which give support to classical mountain-building theories. Evidence for rifting or extensional tectonics are also present.
14	Eastern and Western Atlantic	Marine sediments were deposited prior to separation of South America and Africa. Shallow-water fossils found in deep-water deposits on both margins of Atlantic as result of transportation by turbidity currents.

15	Caribbean Sea	For the first time, strata bridging the interval from late Paleocene to early Eocene have been recovered in the Caribbean area.
16	Panama Basin: Eastern Equatorial Pacific	Data confirm the estimates of spreading rates provided by Leg 9 sites and earlier estimates based on ages of surface cores.
17	Central Basin: Pacific Ocean	Evidence of major environmental change of global extent. Data suggest that many parts of the ridge were above sea level during middle Cretaceous.
18	Continental Margin off Oregon and Kodiak Island	Studied and confirmed subduction on both sides of the local ridges along the continental margins. Investigated the composition and age of cold-water microfauna and flora.
19	Bering Sea: North Pacific Ocean	Sediment found as old as late Miocene. Major increase in volcanic activity began about 3×10^6 years ago; may have caused opening of Gulf of California.
20	Western Pacific	Sea floor of Northwestern Pacific is covered by five stratigraphic units. The history is the result of normal sedimentary sequences. Unusually thick sections of chalk and chert are interpreted as records of previous passages of the oceanic crust beneath the equator.
21	Pacific Ocean between New Zealand and New Guinea	Studied a major regional unconformity covering a wide geographic area. It is thought that the discontinuity may be related to the separation and migration of Australia away from Antarctica as well as later tectonics of this region.
22	Eastern Indian Ocean	Region no older than Cretaceous. Sites on the ridge moved rapidly north during this epoch. Sites on either side of ridge showed little or no movement since early Tertiary. Basal sediments get older to the north. Ridge reached sea level early in its history and has subsided as it migrated northward.

23	(a) Arabian Sea (b) Red Sea	Complex pattern of sea-floor spreading. Evidence of reactivated fracture zone during early and middle Miocene time. Continuous sedimentation over much of the area with no large breaks in sedimentation. History of sea-floor spreading in at least two phases. Presence of rich supplies of copper, zinc, and vanadium suggests a potential resource of these metals.
24	Western Indian Ocean	Resolved problems regarding ocean initiation, time and direction of movement of aseismic and seismically active ridges, and reassembly of the microcontinents of the western Indian ocean.
25	Western Indian Ocean	Determined age of basement sediments and interpreted magnetic anomalies, interpreted tectonic history, studied composition of area, and developed biostratigraphic successions for western Indian Ocean.

4-6 The worldwide pattern of sea-floor spreading based upon magnetic and seismic data.

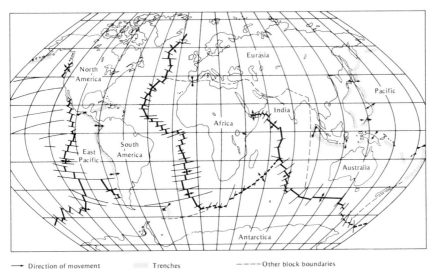

→ Direction of movement   ▨ Trenches   ---- Other block boundaries

── Fracture zones   ▬ Mid-ocean ridges

SEA-FLOOR SPREADING

The theory of sea-floor spreading resulted from the long debate over continental drift. Geologists investigating the shape, structure, and paleontology of the continents recognized from the paleomagnetic data they were collecting that the continents have drifted to their present locations over the expanse of geologic time, but they were uncertain what forces caused such huge landmasses to move over the surface of the Earth.

Oceanographers realized that the floors of the oceans were not flat, featureless plains but consisted of extensive geologic relief. The most prominent feature is the ridge-rift system, which is the extensive seismic ridge found in all the world's ocean basins. Thus began a period of intensive examination. Studies of the sediments collected from the sea floor also led to an interesting discovery, when geologists found that the sediments of the ocean floor were all younger than the Cretaceous period, which began approximately 135 million years ago.

On the basis of these data marine geologists proposed the theory that the floors of the oceans are moving by means of convection currents deep in the mantle of the Earth. This has the effect of bringing material from deep inside the Earth. The molten rock material is released at the axis of the mid-ocean ridges. The extruded rock material spreads out onto the ocean floor.

Additional evidence to support the principle of sea-floor spreading lies in the magnetic particles trapped in the rocks of the sea floor. When the molten rock material being released through the mid-ocean ridges began to cool, the magnetic particles lined up in the direction of the Earth's magnetic pole at the time. Thus, the particles are fixed in that position as the molten material solidifies. The material moving away from the ridges forms bands of rock containing the identifying alignment based on the position of the poles at the time of solidification. A magnetometer (Fig. 4-7) at the surface can then record the relative positions of the magnetized particles. These magnetized bodies also have provided a complete history of magnetic-field reversals extending back to the Cretaceous period.

The history of reversals led to the determination of the rate of spreading in the various oceans. Although the rates differed for different parts of the ridge, the rate of spreading appears to have been constant. The rate of spreading varies from about 1 centimeter (½ inch) to a little more than 9 centimeters (3 inches) per year away from the axis of spreading. This may seem like an insignificant rate of movement, but over the expanse of geologic time the magnitude and effect become impressive.

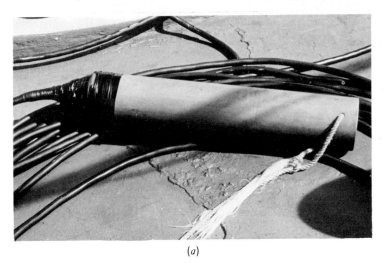

(a)

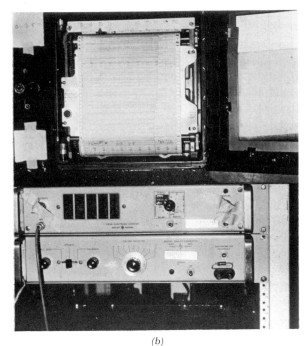

(b)

4-7 A magnetometer fish (a) and recording station (b). (Lamont-Doherty Geological Observatory)

The rate of movement not only appears to be constant, but the motion also takes place in the right direction to account for continental drift. This mechanism particularly supports the concept of two major continents, Gondwanaland in the south and Laurasia in the north. The magnetic lineations mark the path the continents have followed in reaching their present position.

Sea-floor spreading has also been helpful in explaining the absence of large quantities of sediments on the sea floor. As the convection current in the mantle carries the rock material from the interior of the Earth to the ridge crests, it forms new oceanic crust. The new crust is transported on the backs of the convection currents as well.

If a ridge begins to form beneath a continent, the continent will eventually be split and the ridge subsequently located between the two new continents. An examination of the existing mid-ocean ridge reveals that such is the case in the Mid-Atlantic Ridge. If the mid-ocean ridge does not develop under a continent, it will not be located centrally between the continents except by chance. Such is the case of the East Pacific Rise, where the North American continent, driven by a convection current beneath the Atlantic Ocean, is slowly overcoming the ridge and may eventually ride over it completely.

A DYNAMIC EARTH

One of the most recent hypotheses to be presented in respect to a dynamic Earth is the concept of *global tectonics*. In this theory, the origin and evolution of the ocean basin is considered to have resulted when several large plates moved across the surface of the Earth as rigid blocks (Fig. 4-8). According to the data accumulated by Robert S. Dietz and John C. Holden, the Earth, which is composed of an extremely strong lithosphere, or outer shell of rock, also consists of a weaker layer of rock underlying this lithosphere. By the normal forces within the Earth this weaker section has been broken into a series of plates. The continents resting on these plates move across the globe as the plates are set in motion.

The crustal plates correspond in most cases (but not all) to the shape of the continents. The mechanism causing movement is not yet clear, but several possible conditions are suggested. The plates may be pushed, pulled, or carried by the convection currents in the mantle of the Earth (see Fig. 4-9).

In attempting to solve this puzzle, Dietz and Holden have constructed a series of maps to trace the 225-million-year journey of the continents across the face of the Earth. They view the continents during the Permian period, approximately 200 million years ago, as being joined in one massive supercontinent, the Pangaea of Wegener's theory. Pangaea was an irregular landmass surrounded by a single ocean, Panthalassa, the ancestral Pacific. At this time, the continents as we know them today were joined together and located south and east of their present position.

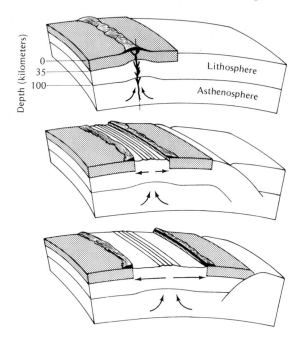

4-8   The formation of tectonic plates due to convection currents within the asthenosphere, the soft layer of the Earth's mantle.

The forces that eventually broke up the supercontinent of Pangaea began during the Triassic era, some 200 million years ago. Two extensive rifts occurred, which separated the continent and opened the Atlantic and Indian Oceans. The northern rift split Pangaea from east to west, creating Laurasia, a landmass composed of what is presently North America and Eurasia. The southern rift split South America and Africa away from the remainder of Gondwana, all of which consisted of what is today South America, Africa, Antarctica, Australia, and India. Shortly thereafter a smaller rift separated India, freeing it to begin its northward journey.

By the end of the Jurassic era, 135 million years ago, the continents consisted of four large landmasses with a rift beginning to split South America away from Africa. In addition, the Atlantic and Indian Oceans were widened considerably.

Then, 65 million years ago, at the end of the Cretaceous period, the rupture of South America and Africa was completed. Australia tore away from Antarctica, Africa drifted toward the north, and India reached the equator. At this stage all the continents as we know them today took form, except for the connections still remaining between

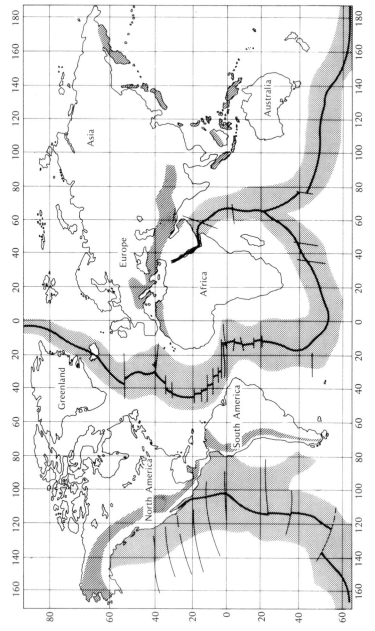

4-9 Worldwide geological activity associated with mountain ranges, earthquake zones, and the trench system.

Greenland and northern Europe and between Australia and Antarctica.

The breakup of the continents was completed during the Cenozoic period, which began about 65 million years ago. At this time, Greenland finally became detached from Europe, and North and South America became joined by the Isthmus of Panama, which was created by volcanic extrusions and the subsequent uplifting of the mantle. India completed its northward journey and collided with Asia (Fig. 4-10), resulting in an upheaval that produced the Himalayan Mountains. Finally, Australia was separated from Antarctica and drifted northward to its present position.

Assuming that Antarctica has remained relatively fixed throughout the period of continental drifting, it is possible to reconstruct the movements of the continents in time and space. North and South America have drifted toward the west, with North America moving more west-northwest. India and Australia were transported toward the north, and Africa and Europe rotated 20° away from each other. India followed a northerly path until its collision with Asia.

4-10 The collision of India and Asia.

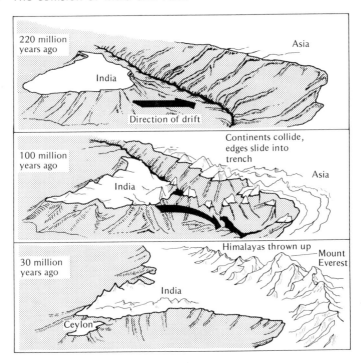

The mechanism of continental drift appears to be the result of magmatic fluids deep within the Earth that well up to form a rift in the ocean floor, causing it to spread. Since the Earth is in a steady state with respect to its volume and total surface area, new crust material coming up along the rift zone must be offset by an equal amount of resorption of that material elsewhere on the surface of the sea floor, perhaps in the deep ocean trenches. No new landmasses are being formed, but the existing ones are subjected to pressures which cause them to drift from place to place.

Relying on their theory of continental drift, Dietz and Holden have extrapolated present plate movements to indicate the positions of the continents in the future. Antarctica will remain relatively fixed, with the Atlantic and Indian Oceans continuing to grow at the expense of the Pacific. Australia appears to be heading northward toward Japan. A rift valley extending along the eastern part of Africa could eventually split off the eastern part of the continent. The remainder of the continent would drift toward the north, closing off the Mediterranean. In the Caribbean, new land areas would develop as the result of compressional uplift. A portion of California west of the San Andreas fault will separate from North America and begin to drift to the northwest. Dietz and Holden predict that Los Angeles will parallel San Francisco in about 10 million years and will slide into the Aleutian Trench in about 60 million years. Perhaps this description will represent the world 50 million years from now.

RESULTS OF THE JOIDES EXPEDITION

The samples collected from the floors of the Atlantic and Pacific Oceans by the ambitious JOIDES deep-sea drilling project will go a long way in helping us to confirm the current theories of ocean-basin origin. Drill cores of the ocean floor reveal a history of the oceans in the collected sediments. For example, sediments collected in the Atlantic indicated that sediments are younger near the crest of the Mid-Atlantic Ridge and become progressively older toward the continents. In addition, the eastern margin of the ocean appears to be younger than the western margin. The change in age of sediments and the calculated rate of spreading is similar to the rate determined by previous investigations of the magnetic anomalies. However, if the continents were once joined together, the age of the crust of the two continents should be the same. A detailed comparison of sediments collected off the continental margin of West Africa contained fossils that were 110 millions years old. This would indicate a difference of 45 million years when compared to the older sediments collected off the continental margins of the eastern United States. The JOIDES scientists attempt to explain this difference by suggesting

(a)

(b)

4-11   Cores taken from sediments are first sampled aboard ship. Note the Nansen bottle in the background. *(a) (Scripps Institution of Oceanography) (b) (Photograph published with the permission of the Woods Hole Oceanographic Institution)*

that the now mighty Atlantic was at one time a narrow ancestral ocean separating the ancient continents of North America and Africa. The two continents perhaps were never fully connected but instead were separated by this narrow ocean, formed as the result of a series of crustal riftings which have since become obscure. The initial rift, independent of the forces causing sea-floor spreading, probably occurred closer to the African continent than the mid ocean. Subsequent sea-floor spreading throughout geologic time enlarged the Atlantic into the ocean we know today.

Preliminary findings of the JOIDES expedition have been made from the examination of drill cores (see Fig. 4-11) recovered during numerous voyages of the *Glomar Challenger* (see Fig. 4-12). In the Mediterranean Sea for example, it has been shown that the drift of Europe and Africa has been stable during the last 130 million years.

4-12 R.V. *Glomar Challenger,* 10,550 tons, 400 feet long, 65-foot beam, operated by Scripps, is the vessel used in the JOIDES program. Visible amidships is the 142-foot derrick. *(Scripps Institution of Oceanography)*

At one time in fact, the Mediterranean Sea was completely cut off from the Atlantic Ocean. Today it appears that a slow drift is bringing Africa and Europe together in the Mediterranean region.

The Geology of the Caribbean Sea Basin

The Caribbean Sea also presents a geologic puzzle. The sea-floor rock in this region is unlike that formed in the adjacent Atlantic Ocean, but there are some similarities between this sea and the Pacific Ocean. The analysis of the Caribbean rock reveals that it is intermediate between rocks formed over the continental landmasses and those typical of the sea floor. Preliminary JOIDES findings estimate that the Caribbean is a relatively young basin formed as a dropped block of rock when North America and South America tore apart during the initial stages of drift.

Curiously enough, the basin adjacent to the Caribbean, the Gulf of Mexico, is an old basin relative to the Caribbean. In this inland arm of the sea sediments have been collecting since the late Cretaceous period, approximately 70 million years ago. Thus, this basin formed rather early in the process of continental drift while the Caribbean formed as North and South America later began to move apart in a north and south direction.

As the JOIDES project continues, engineering developments are providing many new tools and techniques to facilitate the collection of data and samples from the ocean floor. For example, a recent engineering breakthrough now allows scientists not only to replace worn out drill bits on deep-sea drills but also to set the drill back in the same hole on the ocean bottom. In the past if the bits wore out, the drill core had to be abandoned since it was impossible to reenter the same hole. Thus cores now can be recovered from greater depths in the ocean floor.

Better-equipped research vessels add to our knowledge by allowing scientists to analyze the sea-floor sediments aboard ship. This is particularly important in the critical area of marine geochemistry, where changes in the sediments take place rapidly. Unless sediments are studied as they are collected, rapid changes alter them from their original form.

Future investigations will add to the volume of knowledge, but already the data collected by the various JOIDES expeditions strongly support the theory of global tectonics.

SUMMARY QUESTIONS

1. How do the apparent ages of the continents and the ocean basins differ? To what may this be attributed?

2. Describe the theory of continental drift as proposed by Wegener.

3. What geologic and oceanographic data seem to support a concept of drifting continents? List them and explain the significance of each.

4. Describe the current hypothesis for sea-floor spreading and the features attributed to the mechanism causing the motion. In what way would sea-floor spreading cause continents to drift?

5. How are magnetic reversals revealed in sea-floor rock? How does this support the idea of global tectonics?

6. Describe the JOIDES project and the information it is responsible for recovering.

7. What evidence suggests the existence of a large supercontinent in the geologic past?

8. What will the future configuration of the continents be if the present pattern of block motion continues?

9. How are rocks dated? What natural processes allow us to determine the age of a rock?

10. What new technique has been developed by JOIDES scientists for obtaining more data from drill cores?

11. How is global tectonics used to explain the formation of mountain ranges?

12. Which continental block has remained relatively fixed over the last 200 million years?

13. Describe how the Mid-Atlantic ridge formed. What were some of the results of its formation?

14. What original landmass broke up to form North America and Eurasia?

15. What are the ages of the oldest sea-floor sediments?

16. Describe how magnetic "stripes" appear in sea-floor rock.

17. From what area of the world did Gondwanaland derive its name?

18. What activity of the magnetic poles is revealed by magnetic anomalies?

19. What type of "match" was achieved by Edward Bullard and his associates? In what manner were continents matched?

20. How does the mechanism that drives the continental plates apart operate?

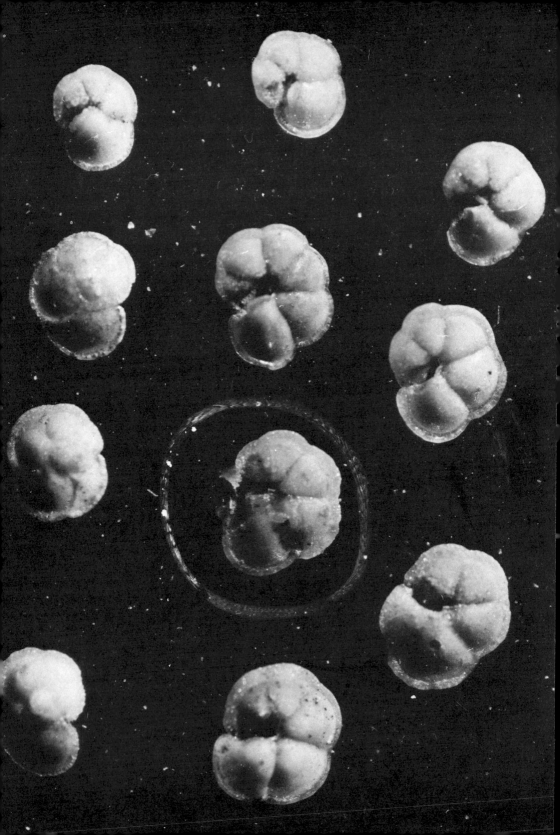

5

OCEANIC SEDIMENTS

The oceans serve as a vast depository for the enormous quantity of sediments eroded from the land and carried by the rivers and streams into the water. These sediments, called *detrital materials,* are carried in many forms, some of which may be organic in origin. In the dissolved state the runoff waters of the Earth carry a significant amount of the salts found in the world's oceans (Fig. 5-1). Insoluble sediments of the Earth are also transported to the oceans. Being denser than the seawater, most of the transported sediments settle toward the ocean bottom when deposited in the sea. However, because of physical and chemical changes, as well as the depth of the sea, many sediments decompose or disintegrate before reaching the bottom. A significant amount of the sediment is derived from the sea itself, in the form of skeletal remains of sea organisms (Fig. 5-2), inorganic precipitates derived from the soluble constituents of ocean water, or products of volcanic extrusions occurring in the sea. A small amount of the sediments can be linked with extraterrestrial material falling to Earth and being deposited in the oceans as meteoric dust.

The ocean floor is covered with sediments. Since sedimentation has been occurring throughout the ages, a study of sediment cores recovered from the ocean bottom yields a picture of past environments and a record of Earth history.

CLASSIFICATION OF SEDIMENTS

Deep-sea sediments may be classified in several ways. The description of the physical

Globerotalia menardii—a microscopic foraminifera. *(Lamont-Doherty Geological Observatory)*

5-1   Continental runoff in the U.S.

properties of the sediments utilizes grain size, color, texture, or composition; and the origin of the sediments is useful in classifying sediments of the deep-sea floor. Although it is often difficult to determine the origin of sediments, one classification defines sediments in terms

5-2   Skeletal structures found in deep-sea sediments.

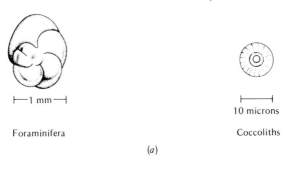

├─1 mm─┤

Foraminifera

├──────┤
10 microns

Coccoliths

(a)

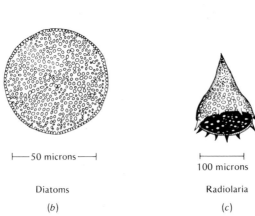

├──50 microns──┤

Diatoms

(b)

├──────────┤
100 microns

Radiolaria

(c)

of origin and mode of deposition: pelagic, originating in the ocean, and terrigenous, or land-derived.

PELAGIC SEDIMENTS

The pelagic sediments of the open sea may be subdivided into organic and inorganic forms. The pelagic deposits are generally light-colored and fine-grained and usually contain some skeletal matter from the myriad of planktonic life-forms that settle to the bottom of the ocean after the organism dies.

Deposits containing greater than 30 percent organic material are referred to as *oozes* or *biogenous sediments*. The chemical composition of the ooze varies. For example, calcareous and siliceous oozes are composed of particles having a large percentage of calcium carbonate or silica, respectively. The oozes are subdivided further by including the name of a characteristic organism found in the sediment. One of the most common calcareous oozes is called *globigerina ooze* after a particular organism. Similarly, a siliceous ooze is called the *radiolarian ooze* because it contains an abundance of this planktonic form.

The inorganic deposits are sometimes referred to as *red clay*. This is a misnomer since very few of the deep-sea sediments are truly red but approximate a deep brown, and brown clay would be a more accurate designation. The inorganic deposits contain less than 30 percent of organic materials, being composed of materials which are low in carbonate and silica content. The brown color of the inorganic sediment is the result of oxidation. The sediments build up at a very slow rate, usually 1 to 2 millimeters per 1,000 years, although certain areas of the oceans experience higher or lower rates. This slow rate of deposition is due to the small size of the particles and the low settling velocity of the sediments. Most of the inorganic deposits consist of very fine-grained, well-sorted muds, made up of the minute detrital particles settling from the upper waters. Where the waters are very deep, only the most resistant particles will ever reach the ocean bottom. Materials that are able to withstand the corrosive agents in seawater reach the floor to become part of the vast sedimentary cover of the ocean basins. The long time required for the very small particles to settle helps disperse the sediments widely. The fine sediments drift for long distances across the sea as they are carried by ocean currents. The inorganic pelagic sediments cover a vast area but are limited to the very deep sections of the oceans. Here the chemical and physical action of seawater has removed the organic materials, leaving only the inorganic forms to be deposited. While the pelagic sediments are widespread in the deepest part of the oceans far from land, the land-derived, or terrigenous, sediments cover a small area of the ocean basin close to shore (see Fig. 5-3).

TERRIGENOUS SEDIMENTS

The terrigenous sediments are derived primarily from the breakdown of continental rock materials. The detrital particles are carried to the ocean by the rivers and streams of the land and deposited nearest the shoreline. The classification of terrigenous sediments, which is more difficult than that of pelagic sediments, is made on the basis of color, texture, and composition. Terrigenous sediments are quite different from the brown clays of the pelagic sediments and vary widely in color from place to place. Table 5-1 gives an idea of the variety of particle sizes and compositions.

The color of terrigenous sediments varies according to the location and the source of the parent material. Generally the color is indicative of the composition of the sediments or the environment where

5-3 Bottom photograph of the eastern Mediterranean Sea showing rock outcrop and erosion of sediments. *(Texas Instruments, Inc.)*

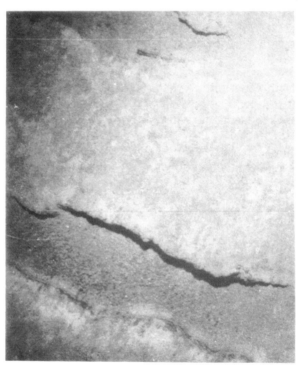

TABLE 5–1   Wentworth Scale of Particle Sizes for Sediments

Wentworth scale		To get next larger size, multiply by:	Approximate equivalent
Size in millimeters	Fragment		In inches
256	Boulder		10
64	Cobble	4	2½
4	Pebble	16	5/32
2	Granule	2	5/64
1/16	Sand	32	.0025
1/256	Silt ⎱ Dust	16	⎧ .06 mm
	Clay ⎰		⎨ .00015
			⎩ .004

SOURCE: Modified after C. K. Wentworth, *J. Geol.*, vol. 30, p. 381, 1922.

they were formed. Red muds, as in the pelagic sediments, indicate the presence of iron oxide. White muds are high in calcium carbonate, blue muds are high in organic material, and black muds form when the organic material decomposes, forming sulfides. This is often evidenced by the strong odor of hydrogen sulfide gas when the sediments are disturbed. The color of terrigenous sediment is so dependent on local environment that it may vary considerably in a very short distance.

Terrigenous deposits are generally coarser than the pelagic forms although some terrigenous sediments are finer than those found in the open sea. Usually limited to dispersal on the continental shelf, the terrigenous sediments are representative of the rocks of the Earth's crust that have been weathered and transported to the ocean by rivers and streams. Most terrigenous deposits are composed of silicates, often containing quartz and feldspar minerals.

The composition of terrigenous sediments is related to the source of the material. It is possible to classify terrigenous sediments on the basis of organic and inorganic constituents. A term such as *coral* may be used to describe the sediments when the remains of the organism coral are dominant in the sediments. If the organic sediments are not dominated by a specific organism but a type, it is possible to refer to those sediments as calcareous or siliceous as an indicator of the kind of skeletal material present.

The inorganic sediments may also be referred to according to the method of transportation responsible for their reaching the areas of deposition. For this reason, inorganic sediments are called volcanic if they contain large amounts of material derived from the eruption of volcanoes. The most common materials derived from this source are pumice, obsidian, and volcanic ash. The same reasoning applies to glacial sediments, which contain large amounts of material deposited by glaciers and icebergs (see Fig. 5-4).

SEDIMENTARY DISTRIBUTION IN THE OCEANS

As the method of collecting bottom samples improves, more and more knowledge of the rate and type of sediments distributed on the ocean floor is becoming available. In the formative years of oceanographic research scientists relied upon fairly unsophisticated devices to collect sediment samples. The earliest method utilized an adhesive applied to the lead weighting sounding lines. Usually some sediments would adhere to the adhesive and be retained as the lead was raised to the vessel. This hit-or-miss proposition yielded little sediment to study and hence little information about distribution of the sediments. As methods improved and the scientists progressed from the various types of grab samplers and corers, knowledge of sediment distribution grew. Although the studies are far from complete, a number of generalizations can now be stated.

As we have pointed out, the pelagic sediments are restricted to the large ocean basins, the most common form of sediment being the globigerina ooze. Clays are also significant in terms of general distribution of the pelagic sediments. The diatomaceous oozes are found in the equatorial and extreme North Pacific and in a belt that encircles Antarctica. Radiolarian oozes are located in a rather narrow belt just north of the equator in the eastern part of the Pacific Ocean. Pteropod oozes, containing shells of pelagic mollusks, are limited to a zone in the Atlantic Ocean located on the Mid-Atlantic Ridge at about 20° south latitude.

The terrigenous sediments are distributed in a wide and irregular pattern. Although the depth of the sediments varies from place to place, depending upon the availability of source material and the

---

5-4 A core prepared for analysis according to depth. *(Lamont-Doherty Geological Observatory)*

depth of the ocean in the area of deposition, it has been known for some time that the greatest deposition occurs in the higher latitudes. For example, the coasts of the north polar basin abound in terrigenous sediments. While the sediments of the high latitudes are composed mostly of mineral particles, the lower-latitude sediments are composed of calcareous remains. In general, distinctions between bands of sediment are obvious over wide areas even though overlapping occurs where the topography is irregular in the marginal areas.

Although our discussion has been based largely on the supposition that the sediments of the sea floor have accumulated as the direct result of settling of the sediments distributed in seawater, oceanographers recognize that some deposits have formed where they are found. These *authigenic* deposits, as they are called, are usually precipitated from the ocean water and undergo some physical or chemical change. The process occurs after deposition; however, changes in the sediments take place before they are buried. The alteration of sediments after they have accumulated on the sea floor is not an example of the formation of authigenic deposits.

Manganese nodules are examples of authigenic deposits on the ocean floor (Fig. 5-5). Manganese nodules are an important potential economic resource because the amount of manganese found in mineral deposits on the continents is limited. These nodules on the ocean bottom are revealed by underwater photography and dredge samples of the bottom deposits. The development of a practical method of collecting the nodules will yield a valuable resource for man. The nodules form when the manganese and iron oxides from the continents form an accumulation on other deep-sea sediments and continue to grow as coat after coat of material is added. The manganese and iron oxide may also be derived directly from the seawater as the result of a chemical reaction of the seawater with volcanic sediments in the ocean.

How fast the nodules grow varies from place to place and is related to environmental factors. However, even though some recent material has been covered with several inches of material, radioactive dating indicates that the rate of buildup in the deep sea is less than 5 millimeters per million years.

Other authigenic precipitates include calcium, carbon, silicon, and phosphorus, as well as many trace elements. These elements are carried to the sea in solution and may be precipitated by a variety of processes. Some of these precipitates may also have future economic significance. For example, phosphorus is important in the production of phosphorite nodules, a necessary element of most fertilizers. Although phosphorite generally is not found in the deep

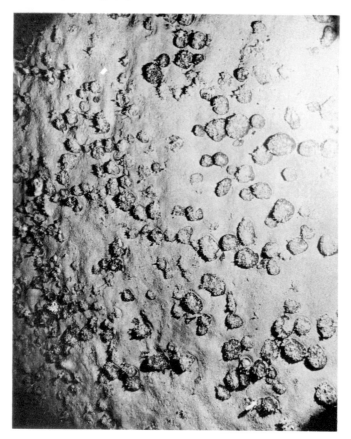

5-5   Manganese nodules.

---

sea, large deposits have been located at a depth of 500 meters (1,640 feet) or less, particularly off the coast of southern California (see Fig. 5-6).

Although this discussion has concerned itself with areas of deposition on the ocean floor, there also are areas devoid of sediments because of the presence of strong currents which either prevented the buildup of sediments or removed deposits of younger sediments by erosion.

Studies of ocean sediments have revealed valuable information about the nature of the oceans through long periods of geologic time. Changes in the world's climate and level of the sea took place during the periods of extensive glaciation (Fig. 5-7). These events in the Earth's history are recorded in the sediment cores brought up from the ocean bottom.

5-6   A rock outcrop. Photo taken by R.V. *Vema* at 1,600 fathoms. *(Lamont-Doherty Geological Observatory)*

SEA-LEVEL CHANGES

Changes in sea level on Earth are minimal when viewed on a day-to-day basis, but over the expanse of geologic time, the variation in the level of the sea has been significant (Fig. 5-8). At any given location on the Earth, the sea surface has alternately been raised or lowered; these changes in sea level have occurred as a result of periods of glaciation separated by the interglacial periods.

During the last 2 million years of geologic history, the Earth has been subjected to a series of rapid climatic changes in the physical environment. Glacial ice periodically covered large areas of North America and the northern parts of Europe and Asia. The alternate growth and decay of these ice sheets had the effect of causing a worldwide change in the level of the sea, or *eustacy*. The water to form these glaciers was derived from the oceans, and the buildup of ice lowered the sea level during periods of glaciation. During the interglacial periods the glaciers melted and returned the water to the oceans, causing the level of the sea to rise.

From estimates of the thickness of ice during the Pleistocene epoch

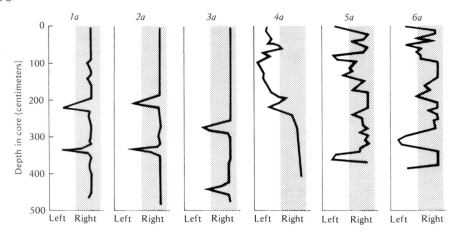

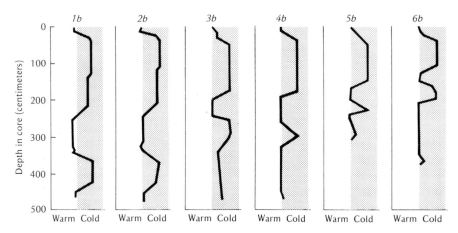

5-7 Climatic shifting variations in the past can be correlated from the characteristics of planktonic life-forms in warm- and cold-water masses.

of glaciation (2 million years ago), it is thought that the level of the sea was more than 100 meters (328 feet) below its present level. If this estimate is correct, most of the present-day continental shelves were exposed at that time. The lowered seas played an important part in forming the continental shelves, which cannot be viewed as extensions of the continental landmass. During this time there were four major lowerings of the sea produced by the four glacial periods (see Fig. 5-9).

Contrary to earlier belief, it is now recognized that the finer sedi-

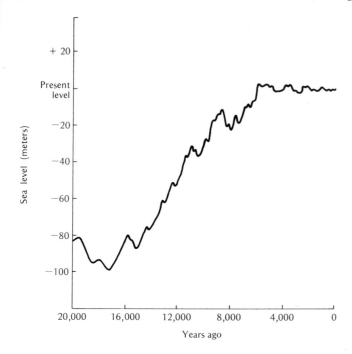

5-8 The changes in sea level indicating a relatively constant sea level over the last 6,000 years.

ments are not necessarily located closer to the continents nor coarser sediments necessarily at the shelf break. Recent studies have shown that the size of the sediment and its distance from shore are unrelated. Since the last period of lowered sea level, some 15,000 years ago, the sediments on 70 percent of the world's continental shelves have been set down. The remaining 30 percent of the shelf area is covered by silts and sands deposited on the broad coastal plain by rivers moving across the shelf to deposit most of their sediments in the deeper parts of the ocean. The coarser sediments near the shelf break may be explained by assuming that with no shelf to reduce their height, waves were probably higher at the shoreline than they are today. This would account for the coarser sediments deposited in that area.

More evidence for a reduced sea level is becoming available as studies of the shelf sediments are completed and reported to the scientific community. Remains of freshwater debris have been sampled in many places around the world. When the sea level was reduced, the lowlands were conducive to the formation of ponds and

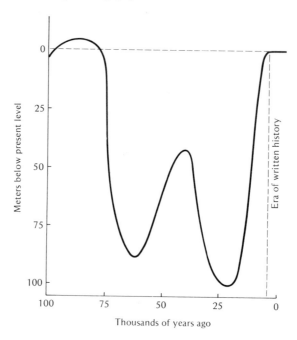

Meters below present level

Era of written history

Thousands of years ago

5-9 Sea-level changes effected by the glacial ages over the last 100,000 years.

marshes that soon filled with the debris of the forests and meadows surrounding this lush environment. It is that debris, now submerged, which suggests the earlier exposure of the shelf areas. In addition, carbon 14 data obtained for a number of deposits collected from the shelf make it possible to depict the changes in sea level graphically. The data show that the sea was approximately at its present level about 35,000 years ago and that it began to recede about 30,000 years ago; 15,000 years before the present the sea level had dropped 130 meters (426 feet) or more, rising rapidly after this time so that it approached within 5 meters (16 feet) of the present level of the sea 5,000 years ago. During the last 5,000 years it has risen to its present level at a slow and irregular rate.

Additional evidence for a recent rise in the level of the sea has come from the remains of animals attracted to the densely vegetated region of the extended coast. Bones and teeth of mammoths and mastodons, land-dwelling animals, have been collected off the coast of the United States, Europe, and Japan. Bones of musk-ox, horse, and moose have also been found on what is now the submerged part of the continental shelf.

Presently, we are experiencing an average rise in the level of the sea of 10 centimeters (4 inches) per century. Given the amount of glacial ice that still remains on Earth, we have not yet reached the ultimate level of the sea. The current warming trend would indicate that we are not completely out of the period of glaciation; if the warming trend continues at its present rate, the level of the sea could rise an additional 50 meters (165 feet) or more. Although this phenomenon will not occur in our lifetime, the effect will be devastating to the large segment of the world's population living near the coast.

Besides the direct effect that the alternate melting and freezing of the ocean water had on reducing or increasing glacial ice and thus on the level of the sea, it is believed that a more complex phenomenon was taking place simultaneously. During the glacial period, huge sheets of ice were formed and concentrated on a relatively small area of land. According to the theory of isostatic change (described in Chap. 3), the buildup of ice on the landmass depressed the crust, uplifting the ocean floor. The glaciated surface sank while the unglaciated surface rose; this could account for a change in the level of the sea as well.

Drowned coral reefs show up on many of the charts produced by echo soundings of the deep ocean. The rate of sea-level rising was too rapid to permit these reefs, which form only in shallow water, to continue to grow; they were eventually covered by sediments. Since the present sea level was reached during the last 5,000 years, modern, living coral reefs must have formed since that time.

CORAL-REEF FORMATION

Coral reefs are biological communities built on the floor of warm shallow waters. Although ancient reefs are found in many parts of the world, the modern reef is limited to the shallow water of the tropics and subtropics. This serves as one more bit of evidence of the changing climate of the oceans throughout geologic time. Although there are a few exceptions, corals grow best in clear water less than 35 meters deep and a water temperature of at least 21°C (70°F). Only colonial forms of coral are capable of building reefs. They build new organisms upon old to extend the colony from deeper water upward toward the surface.

A recent discovery may prove useful in dating changes in sea level: some corals grow by adding layers. In some varieties of corals, this layered growth is a daily event. Although not well defined, some coral rings have been counted, suggesting the future use of coral as a paleontological clock. If so, a coral dated by radioactive means could also be used to determine the number of days in a year at that time in the past. For example, in two cases where the daily rings were visible enough to be counted, the corals were dated by radioactive means and found to be from the Middle Devonian age, about 350 million years ago, and the Pennsylvanian age, about 280 million years ago. The number of daily rings between the annual rings was 400 and 390, respectively, a reduction that closely agrees with the estimated reduction in the length of the year that resulted from the slowing of the Earth's rotation.

The corals considered here are generally referred to as *stone corals* because their skeletons are composed of hard calcium carbonate, which produces a structure strong enough to withstand the pounding force of the waves. The reef community is made up of other organisms besides coral. Calcareous algae are common on the seaward side of the reef, where they grow over the coral, forming a solid structure which strengthens the reef against the onslaught of water. The corals flourish on the reef flat and build up a colony consisting of many branching and massive spherical corals. Dead coral, sand, and other detrital materials fill in the reef flat, to produce a shallow platform on which many delicate corals, sea fans, sea urchins, and sponges survive. The back reef is composed of carbonate sediments of sand-sized particles which are skeletal remains of portions of the main reef which have broken off and been mechanically ground down to form the detrital particles.

Generally, coral reefs take three forms, the simple *fringing reef,* the *barrier reef,* and the ringlike *atoll* (see Fig. 5-10).

The fringing reef grows out from a landmass and always maintains contact with it. A remarkably flattened reef, it varies in width and may be as much as ½ mile wide. A fringing reef borders the Florida keys.

Barrier reefs are long and generally narrow reefs that lie offshore, separated from the land by a lagoon. The offshore breakwaters they form serve to protect large islands and even the coasts of continents by reducing the force of water moving onshore. The largest of this type, the Great Barrier Reef, is over 1,400 miles long, averages 100 miles wide, and lies along the continental shelf off the coast of Australia. Smaller (although still quite large) barrier reefs are found in the Bahamas and around islands in the tropical western Atlantic and South Pacific (see Fig. 5-11).

Atolls, the most common and best-known form of coral reef, are ring-shaped reefs of coral rising from deep water and enclosing a lagoon. They appear in isolated groups, particularly in the Pacific

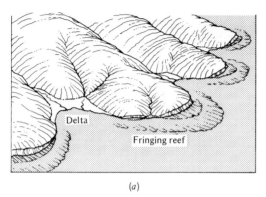

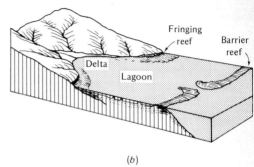

(a)

(b)

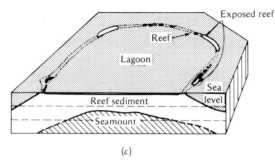

(c)

5-10 Types of coral reefs.

Ocean and are often associated with the formation of seamounts and guyots. Many questions remain unanswered about the origin of atolls, as they do not expose the original foundation on which they are built. Interest in the formation of atolls and barrier reefs can be traced back to Darwin. In 1842 he proposed, as an outgrowth of his voyages on H.M.S. *Beagle,* a hypothesis which was very simple yet essentially correct. It took almost 100 years for scientists to realize the validity of Darwin's theory.

Darwin suggested that, as an initial step, a fringing reef develops during the slow subsidence on the sides of a volcanic island (Fig. 5-12). As the island continues to subside, the coral continues to grow. Eventually, as the corals grow and the landmass continues to subside, the barrier becomes inundated to form a lagoon. As subsidence continues, the island is buried, the lagoon widens, and the reef forms an atoll. Darwin also suggested that not all atolls have to evolve through the barrier-reef stage: it is possible for atolls to form directly.

5-11   A living coral reef.

He even went further, perhaps anticipating the findings of the future, to state that it is possible for coral to develop on a shallow bank and form a structure without subsidence that would be difficult to distinguish from the true atoll. Evidence collected from drillings at several atolls proved Darwin essentially correct in his thesis. Further investigations conducted by the U.S. Geological Survey in 1952 revealed that the Marshall Islands originated (at least in part) on flat banks and were subsequently converted into atolls. These corals probably never passed through the intermediate barrier-reef stage. Thus, Darwin's theory was unquestionably confirmed.

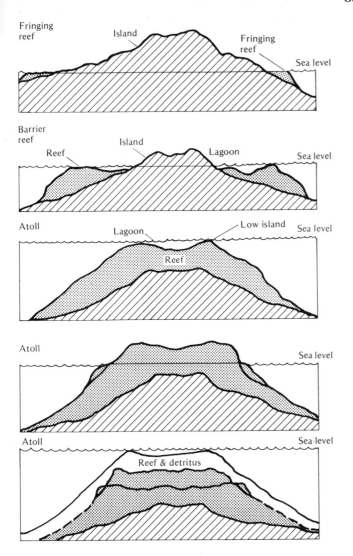

5-12   The development of coral reefs according to Darwin.

SUMMARY QUESTIONS

1. List the two major sources of oceanic sediments and describe four main types of each.

2. What are the origins and sources of brown clay?

3. How are deposits such as manganese nodules formed on the sea floor?

4. How are sediment samples utilized to correlate and determine climatic changes?

5. What evidence have we for past changes in sea level? How did sea-level changes affect the coastlines of our continents?

6. What conclusions did Darwin draw regarding the relationship of coral growth to sea level?

7. Describe the scheme used to classify ocean sediments.

8. What value are the sediments collected from the sea to man's quest for an accurate history of the sea's past?

9. How do we account for the widespread distribution of sea sediment?

10. Is sedimentation taking place at a fixed and constant rate? Why?

11. Describe the various types of coral formations. How are they classified?

12. What are stone corals? How do they form?

13. Describe the growth and evolution of a coral reef.

14. Describe the general characteristics of sea-level changes for the last 100,000 years.

15. How is sea level changing today? Why?

16. What effect did ancient ice sheets have on the earth's crust?

17. How do some plankton remains reveal climatic changes?

18. What does eustacy mean?

19. How do authigenic deposits form?

20. List four sources of sediment on the ocean floor.

21. Describe the general distribution of sediments in the ocean.

22. How was Darwin's theory of coral-reef formation confirmed?

23. What would be the major effect of the present changes occurring in sea level?

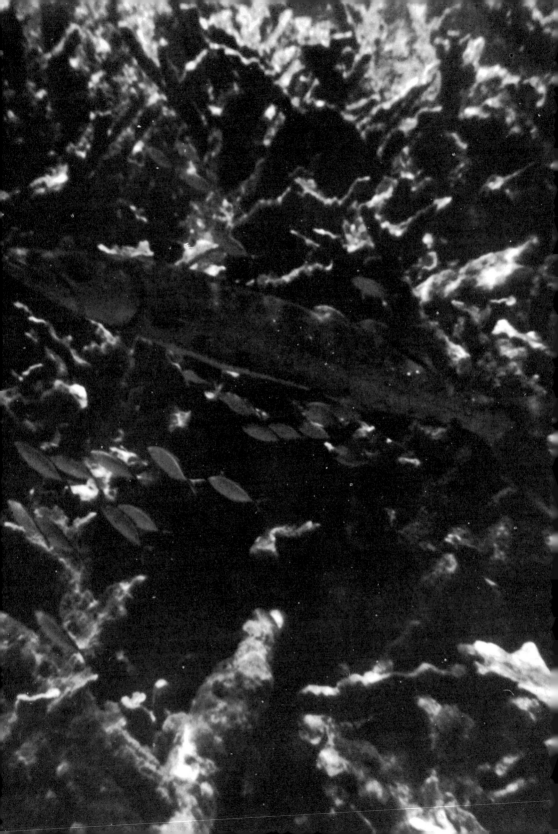

6 | THE WATERS OF THE OCEAN

Water is an ordinary and universal substance—or at least it appears so to the casual observer. However, nothing could be farther from the truth! Water is a complex liquid capable of physical change and possessing the most unusual properties of all the compounds on Earth.

It is thanks to the physical and chemical characteristics of water that life on this planet has evolved into the enormous range of forms found on land and in the sea. Water readily dissolves the many elements and compounds necessary for metabolic reactions and is highly stable under ordinary conditions. Water dissolves many organic and inorganic salts as it flows over the land through stream and river channels and makes its way to the sea.

Water is close to being the universal solvent sought by the ancient alchemists. Given sufficient time to effect its work, water wears away the continents and slowly dissolves vast amounts of minerals. Approximately 50 million tons of material removed from the continents enter the oceans as dissolved salts and suspended particles each year. Vast quantities fall to the floor, accumulating as thick sediments over long periods of time.

THE NATURE OF PURE WATER

Pure water is the basic substance in which the different salts of seawater occur in different proportions to make a variety of solutions. Water, in turn, is affected by local and seasonal conditions. The physical properties of pure water permit the many interactions of energy between the ocean and the

A coral reef in the tropics.

atmosphere that produce our weather, climate, and manifold life in the sea.

Water is composed of two atoms of hydrogen and one atom of oxygen. For many people its formula, H_2O, sums up their total knowledge of water. Yet it is only the starting point for our discussion.

Each hydrogen atom contributes one electron, a negatively charged particle, to the oxygen atom. The oxygen atom achieves a chemical equilibrium as it acquires the electrons contributed by the hydrogens. This bond between atoms creates a molecule of water with the hydrogen atoms aligned near one end of the oxygen atom at an angle of 105° (Fig. 6-1).

As a result of this structure, the water molecule resembles a bar magnet; it has oppositely charged poles at either end, like the north and south poles of a magnet. For this reason water is said to be *bipolar*. The two positively charged hydrogen atoms near one end of the molecule and the double negative charge of the oxygen atom at the other end are responsible for the bipolar nature of the molecule. Consequently, adjacent water molecules tend to join together, held by the attraction between opposite charges on adjoining molecules. As water molecules make contact with one another, the positively charged hydrogen atoms of one molecule are attracted to the negatively charged oxygen atom of the other molecule. This bond of association between molecules is called a *hydrogen bond* (Fig. 6-2). It causes a water molecule to associate with other water molecules more strongly than uncharged molecules associate with each other. This association produces a greater stability between adjacent molecules than one would normally expect and prevents the easy separation of water into individual molecules. Thus water has a highly stable molecular arrangement.

Water is the only substance that can be found in nature in all three states of matter—solid, liquid, and gas—simultaneously under ordinary conditions. If the molecules of water were not held together by the hydrogen bond, it would have a boiling point of about −80°C (−112°F), but because of the great stability of the hydrogen bond,

---

6-1 The configuration of the water molecule.

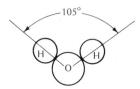

6-2 The hydrogen bond linking adjacent water molecules.

___

the boiling point of water is 100°C (212°F). The melting point of ice is 0°C (32°F); following the same line of reasoning, ice would be expected to melt at a much lower temperature, approximately −100°C (−148°F) (see Fig. 6-3).

WATER AND HEAT ENERGY

Water can absorb greater amounts of heat than most other substances. This quantity, called *specific heat,* is the amount of heat energy required to raise the temperature of a given quantity of a substance by a specific amount. For example, water will absorb a larger amount of heat by 3 to 5 times before its temperature rises to the same degree as soil.

By definition, the amount of heat required to raise one gram of water one degree Celsius is one calorie of heat (the Celsius scale used to be called the centigrade scale). When it is stored in the system, this is known as *latent heat.* It can be released later to the atmosphere

___

6-3 A comparison of the Fahrenheit and Celsius temperature scales.

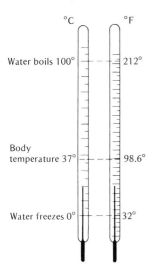

or to cooler bodies of water. The latent heat stored in the sea is an important factor in the exchange of energy that creates weather systems throughout the world (Fig. 6-4). The heat energy exchanged between the oceans and the atmosphere also alters the density of water masses. Thus, heat energy plays a role in oceanic circulation, as discussed in Chap. 7.

Temperature

The temperature of the open sea is determined primarily by the angle at which the sun's rays enter the Earth's atmosphere and the amount of energy reaching the surface. The equatorial regions, where the rays of the sun are direct throughout the year, receive about 3 times as much solar energy as the polar regions, where the rays of the sun strike the Earth at an oblique angle (Fig. 6-5). Thus, the solar energy received at the equator is more concentrated than the energy received at the poles. There is an excess of energy absorbed at the equatorial regions and a deficit of energy in the higher latitudes. The energy balance between all parts of the globe and the total energy received by the Earth is achieved by means of interactions between the circulating sea and atmosphere.

6-4   The phase changes of water.

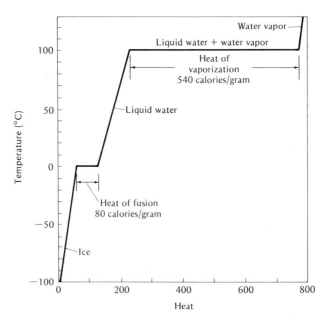

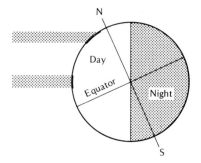

6-5 Insolation variations with latitude.

Most of the solar energy is absorbed within the first 300 feet of the sea surface. Over the open sea, the range in temperature at the surface is from approximately −2 to 30°C (29 to 86°F). Salt water does not usually start to freeze until about −2°C.

However, an examination of cold bottom water would reveal the range of temperatures to be much less than on the surface of the sea. Most bottom water is between −1.8 and 5°C (30 and 40°F). These bottom currents move slowly from polar regions toward the lower latitudes of the equatorial zone.

The circulation of surface currents and slower motion of bottom currents achieve a balance in the total heat received by various latitudes throughout the world. Cold, dense water slowly settles to the bottom, where it moves as bottom currents. Surface currents carry warm water into colder regions and cold water toward the equator.

The Thermocline

A cross-sectional analysis of ocean water reveals that temperatures decrease with depth. In all oceans, there usually is a year-round depth boundary known as the *permanent thermocline* (Fig. 6-6). Temperatures vary only slightly in the upper levels of the sea. Mixing and convection maintain a fairly uniform temperature in the upper layers of water. Although temperature changes can occur at shallow depths, from a maximum depth of approximately 500 meters (1,600 feet) a sudden temperature decrease can be noted which continues to about 1,200 meters (4,000 feet). From this level, a constant cold continues to the bottom. Thus, the surface zone is separated from the colder, denser, deep zone by the thermocline boundary.

Little mixing and convection take place in the lower zone. Here, the water moves slowly as bottom currents, but it does not interact

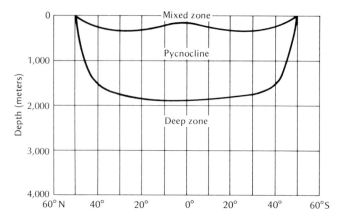

6-6   The main thermocline.

with the upper zones very easily. The convection and overturning in the sea are confined mostly to the isothermal water above the thermocline.

In regions along coastlines, and in shallow bodies of water especially, a number of smaller thermoclines occur, for the most part local in nature. In addition, thermoclines vary, both in number and in depth, with changes in seasons.

Density

Water also exhibits an unusual set of properties in respect to changes in density as it cools and heats.

Like nearly all substances on Earth, water slowly contracts as its temperature drops. As cooling continues, the contraction means a progressively higher density, or mass per unit volume, as the molecules pack tighter together. However, as water continues to contract, the peculiarity of the molecule is such that it reaches its greatest density at approximately 4 °C (39 °F). At this temperature water has achieved its greatest density and as it continues to cool, begins to expand into a larger volume (Fig. 6-7). In the ocean the water mass continues to expand until the crystalline solid (ice) is about one-seventh larger in volume than an equal mass of seawater surrounding the ice (the dissolved salts in the ocean play a role in the density difference as pure ice is actually only one-ninth less dense than *pure* water). The important result is that while colder water progressively sinks downward, ice floats near the surface of the water.

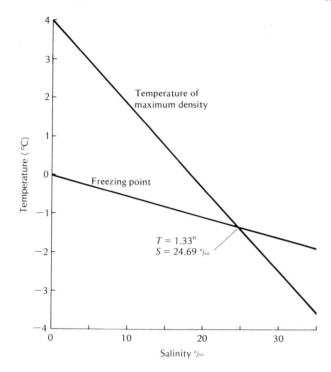

Temperature of maximum density

Freezing point

$T = 1.33°$
$S = 24.69 °/_{oo}$

Temperature (°C)

Salinity °/oo

6-7 The relationship of density to the temperature and salinity of water. (After H. J. McClellan, "Elements of Physical Oceanography," Pergamon Press, 1965)

Imagine the results if water were to freeze from the bottom upward instead of the other way around. As the water slowly changed to ice, in every body of water—lakes, pools, oceans—all life would be slowly eradicated. It is unlikely that the first few forms of life would have been able to succeed on this planet in such circumstances. If the early life-forms existed in pools of water that froze from the bottom upward, the mass of ice undoubtedly would have killed off the primitive organisms and prevented their evolution to more complex life. Since water produces ice masses near the surface, life can continue at a slower metabolic rate beneath, and various forms of living things can survive from one winter to the next.

Although the temperature of ocean water is of major concern, other factors also play a role in determining its characteristics. The composition of ocean water is a result of many factors operating on a worldwide basis.

THE COMPOSITION OF SEAWATER

Ocean water contains a variety of substances dissolved in the water and contained as suspended particles. The variations in composition from place to place depend upon local circumstances, such as the abundance of life-forms, the presence of rivers, and other geological and meteorological conditions. One group of these dissolved materials, the gases, enter the ocean as solutions from land-based streams and from the solutions produced at the water-atmosphere interface. Although many gases are found in seawater, a few are of major importance because of their biological significance.

GASES IN THE SEA

The major gases found in seawater in the order of their relative abundance are nitrogen, oxygen, and carbon dioxide. Others, such as hydrogen sulfide gas, are significant because they indicate bacterial activity, decay of organic materials, and the stagnation of water.

Nitrogen

Approximately 64 percent of the dissolved gases in seawater consists of dissolved nitrogen. This is less than the quantity of nitrogen in the atmosphere, which contains about 78 percent by volume. As in the atmosphere, nitrogen dissolved in seawater is not biologically important, as most creatures cannot utilize the free nitrogen directly. Nitrogen compounds essential to the diet of most creatures are obtained from plant and animal tissue ingested as part of the food chain (see Chap. 13). The free nitrogen must be *fixed* into compounds by a variety of organisms, such as nitrogen-fixing bacteria. The nitrates are produced by chemical reactions carried on during animal and plant metabolism, thus making them available for other life in the sea.

Oxygen

Oxygen occurs in solution as a by-product of photosynthetic plants. It also is dissolved at the interface between the sea surface and the atmosphere. The sea contains a much higher percentage of oxygen (34 percent) than the atmosphere (about 21 percent).

Most of the oxygen-rich ocean water and the animal life which uses oxygen for respiration can be found near the surface. In the first few hundred feet of the sea, most of the overturning and convective mixing occurs, along with photosynthesis carried on by plankton. Thus, most of the available oxygen lies within a few hundred feet of the surface. As one descends into the depths, the amount of oxygen in the water drops rapidly and with it animal life. This oxygen-minimum layer forms a boundary delineating the upper layers of oxygen-rich water from the relatively oxygen-poor layers below (Fig. 6-8). From the oxygen-minimum layer downward, the amount of dissolved oxygen increases initially, and another decrease occurs to the bottom. The presence of oxygen throughout the water does reveal, however, that there is some circulation and interaction between waters of all levels.

Photosynthetic plants require sunlight in order to carry out their activities, and since sunlight does not penetrate ocean water to any great depth, these phytoplankton are found to be concentrated in the upper levels of the sea. As one descends into the ocean, plant life— especially green plants—is almost nonexistent. Thus, the presence of oxygen in greater abundance nearest the surface is a reasonable consequence of the activity of these green plants acting in conjunction with the solution of atmospheric oxygen.

Carbon Dioxide

Carbon dioxide, CO_2, is an important constituent of ocean water. Entering the water as a dissolved gas, it forms a weak acid, carbonic acid, H_2CO_3, in its initial stage of reaction. This gas combines

---

6-8   The oxygen-minimum layer.

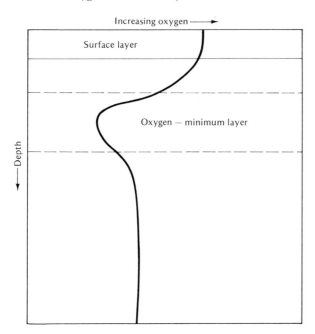

Increasing oxygen ⟶

Surface layer

Oxygen — minimum layer

Depth

with seawater and gives rise to carbonates, which are found in large quantities locked up as rocks, corals, shells, and various sediments.

$$CO_2 + H_2O \rightleftharpoons H_2CO_3$$
$$2H^+ + CO_3^{--} \rightleftharpoons HCO_3^- + H^+$$

Carbon dioxide, required for photosynthesis, is used in large amounts by the abundant green plants of the ocean. Nevertheless, the dissolved carbon dioxide in ocean water (approximately 1.6 percent of all gases) is about 50 times the ratio of this gas in the atmosphere. Thus, seawater absorbs and holds large quantities of carbon dioxide. As carbon dioxide is removed from the water by green plants during photosynthesis, carbonate and bicarbonate compounds are formed which later may release carbon dioxide through chemical action.

Hydrogen Sulfide

All the remaining gases dissolved in the sea make up only about 0.5 percent of the total amount of gases in the water. Of all these, only hydrogen sulfide gas, H_2S, requires mention here.

High concentrations of hydrogen sulfide in water places larger life-forms in danger. Hydrogen sulfide gas is a waste product of the bacterial decay of organic compounds.

In stagnant water or on areas suddenly invaded by warm, oxygen-poor water, bacterial action attacks the available organic compounds. As the bacterial processes of decay further deplete existing supplies of oxygen, death results for the creatures remaining. Thus, the presence of hydrogen sulfide in large amounts indicates that nearly all the dissolved oxygen gas in the water body has been depleted.

Warm water cannot hold as much oxygen in solution as cold water. In areas where upwelling of cool water occurs (Fig. 6-9), such as the west coast of Peru, the oxygen- and nutrient-rich bottom water reaching the surface supports a large fish population (see Chap. 8). During certain seasons, as upwelling ceases and warmer equatorial water invades the area, the nutrient-rich bottom water is not available to support large numbers of fish. The resulting hydrogen sulfide gas enters the air and blackens the white lead-base paint on houses and ships. The people of the region have named the phenomenon the "Callao painter," after a local port city.

Thus, both the ocean and the atmosphere store gases. The ocean absorbs excess gases from the atmosphere. It also releases gases into the atmosphere as they are utilized by land processes, and their concentration decreases in the atmosphere. The ocean and atmos-

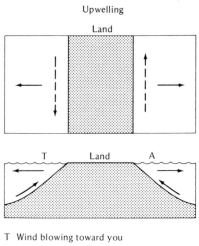

Upwelling

Land

T Land A

T Wind blowing toward you

A Wind blowing away from you

⟶ Water movement

| Winds

6-9 Upwelling as it is induced by prevailing winds.

phere act together until an equilibrium is achieved between the percentage of gases held by the atmosphere and absorbed by the ocean. Although the percentages change periodically, and with long periods of time, the actual amounts of gas in each reservoir are extremely stable and are affected only by changes in basic, fundamental processes throughout the world.

THE MINERAL CONSTITUENTS OF SEAWATER

The sea contains a large number of dissolved compounds and elements. Despite local variations within this constituent mass of material, the percentages of dissolved salts also have remained remarkably constant throughout the world and during the passage of geologic time.

In addition to the 10 major elements found in seawater, at least 49 minor and trace elements are present (see Table 6-1). Many of the minor constituents of seawater are important in the nutrition of various marine life-forms, among them being nitrogen, phosphorus, silicon, potassium, copper, and zinc. The first three elements are present in the water in minute amounts, and the necessity for some of

TABLE 6–1   The Ten Major Elements of Seawater

Element	Symbol	Average in parts per thousand
Chlorine	Cl	19.35
Sodium	Na	10.76
Magnesium	Mg	1.29
Sulfur	S	0.88
Calcium	Ca	0.41
Potassium	K	0.38
Bromine	Br	0.06
Carbon	C	0.03
Strontium	Sr	0.01
Boron	B	0.005

these elements is still unclear to biologists. But tests reveal that without the presence of some of these trace elements, marine life does not flourish.

Most trace elements—and the major elements to a more limited extent—are easily depleted by marine life-forms in the upper levels of the ocean. As rapid plant growth occurs, the elements are used up and find their way to the lower levels in detritus from the dead life-forms which settle toward the bottom. The problem of rapid depletion does not generally exist for chlorine, sodium, magnesium, nitrogen, and phosphorus, as they are present in abundance. Minor elements utilized by bottom dwellers slowly work their way through the food chain and return to the water by subsequent processes of decay.

Most dissolved elements in the sea are found in ionic form. These sea salts, although found in a variety of combinations, consist primarily of a few major types. The majority of ionic sea salts result from the following compounds: sodium chloride, $NaCl$; magnesium chloride, $MgCl_2$; magnesium sulfate, $MgSO_4$; calcium sulfate, $CaSO_4$; potassium sulfate, K_2SO_4; magnesium bromide, $MgBr_2$; calcium carbonate, $CaCO_3$; sodium sulfate, Na_2SO_4; and potassium chloride, KCl.

Many of these compounds are essential to biological processes in marine plants and animals. Although the list is only a representative sample, seawater contains so many compounds that it is one of the most complex solutions on our planet.

THE DETERMINATION OF SEA SALTS
A number of methods are used to analyze the composition of seawater. Although the processes test for various properties, both

chemical and physical analyses of seawater generally involve a measurement of the quantity of salts present in a specific volume of water.

Salinity

The salinity, or salt content of seawater, can be calculated by chemical analysis of the water, but it is generally determined most easily and accurately by measuring the electric conductivity of the water. Pure water is not a conductor by itself, but dissolved salts ionize in the water, forming a solution containing charged particles. The ionized salts carry an electric current, and the amount of dissolved salts controls the amount of current that passes through the water.

The precise definition of salinity is "the total amount of solid material dissolved in a kilogram of seawater when all the carbonate has been converted to oxide, all bromine and iodine have been replaced by chlorine, and all organic matter has been completely oxidized." For our purposes salinity will be "the weight of dissolved salts in a mass of seawater expressed in parts per thousand."

Nearly all the 92 naturally occurring elements are found in the dissolved salt compounds in seawater. An average salinity figure for all the oceans of the world is 3.5 percent by weight. This average is expressed as 35 parts per thousand. The actual mean salinity for all the oceans is 34.7 parts per thousand.

Factors Affecting Salinity

The salinity of the world's water bodies varies somewhat from the average but only over a narrow range in the light of total percentage. In the Arctic Sea, where the area receives much freshwater from land rivers and a low rate of evaporation exists, the average salinity is about 33 parts per thousand. In subtropical seas, where evaporation is high and few rivers dilute the ocean water, salinity is about 37 parts per thousand. At the extreme, one can find a region such as the Red Sea, which has a high rate of evaporation and no freshwater entering the water body, with a salinity of 41 parts per thousand. Thus, the range of salinity throughout the oceans is rather small. Yet the variation determines the survival of many species of life in the sea. The adaptation of marine life is, in many cases, quite precise, and variations in salinity are fatal.

The Halocline

The salinity of seawater varies with depth as well as latitude. As salinity increases, density increases, and water with a high salinity will tend to sink to a level that achieves an equilibrium. Thus, surface water in an ocean has a uniform salinity to a depth of a few hundred meters. On the average, at a depth of about 500 to 600 meters there is a sudden increase in salinity at a boundary called the *halocline*. This increased salinity continues for several hundred feet and then decreases very slowly to the bottom (Fig. 6-10).

THE ORIGIN OF THE CONSTITUENTS OF SEAWATER

The temperature of water affects the amount of dissolved materials contained in it. In addition, the mixing and overturning of the water also determine the amount and type of materials dissolved in the water. The sources of the constituent elements and compounds are an important consideration in recreating the evolution and development of the oceans.

The salinity of the sea has varied somewhat throughout time, but contrary to what is popularly believed, the sea probably did not evolve from a large body of freshwater into a progressively saltier one. The constituent elements of salinity have been constant for a long time.

The salt content of the present sea did not accumulate from a simple repetitive dissolution of materials by rain running over the continents as streams and rivulets. The salt has been contributed by many sources, and the water itself has not come exclusively from rainwater. Careful calculation has revealed that the primitive atmos-

6-10   The halocline; the thermocline.

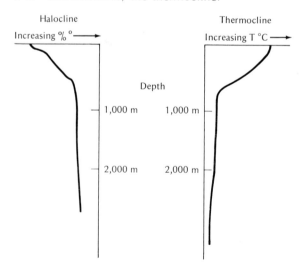

phere did not contain enough water vapor to have formed the vast world ocean.

Much of the present water on our planet was probably released by volcanic eruptions as *juvenile water,* that is, water that had not previously taken part in the cycle of precipitation and evaporation. Along with this water vapor the volcanoes also yielded chlorine, bromine, carbonates, sulfurous fumes, and vast amounts of other gaseous products.

The metallic elements now in the sea undoubtedly did reach it as dissolved products of weathering processes. Potassium, sodium, calcium, magnesium, and other constituents were eroded from rocks, chiefly igneous rocks, by chemical weathering processes and running water.

Some elements entered the sea from terrestrial sources or were obtained from the decomposition of compounds already in the sea by photosynthetic plants. Bacteria break down organic compounds in deeper water and return elements to the water as ionic particles, thus enriching bottom waters. Trace elements are concentrated by planktonic life-forms, and the mixing seawater carries the elements throughout the depths of the sea.

Finally, interactions with the atmosphere form an important part of the resource pool of the sea. Gases and particulate matter are dissolved in seawater at the ocean-atmosphere interface. Further, diatomic oxygen, O_2, is broken down and released as monatomic oxygen, O.

OTHER PROPERTIES OF SEAWATER

In modern oceanography, the sea life, sea floor, and seawater are analyzed by a variety of techniques. Because we now use photography and electronics, we are interested in how water in the sea affects the transmission of light and sound. Salinity, temperature, and pressure all enter into such studies.

The transmission of light is unaffected by salinity, temperature, or pressure. The salts dissolved in the sea do not affect the passage of light through the water, but particles of solid material suspended in water scatter the light.

The usual blue color associated with the open sea is generally considered to be the result of molecular scattering of the wavelengths of light energy entering the water. The wavelengths of blue light are the most penetrating and thus scattered the most. The variety of color frequently noted along coastlines is probably the result of pigments from photosynthetic plankton and suspended particles.

In estuaries and semienclosed bays the water frequently has a brownish cast as the result of organisms and detritus suspended in the water.

Sound waves are easily affected by slight changes in the character of seawater. The speed of sound waves increases with an increase in salinity, temperature, or pressure. In addition, the sound waves may be bent, or refracted, by changes in any one of these.

Sound waves generally cover a speed range from about 1,400 to 1,550 meters per second in water. The average speed of sound in ocean water is 5 times the speed of sound in air, which is about 1,330 meters per second.

As temperature increases, the speed of sound increases at a rate of 3 meters per second per degree Celsius increase. An increase in pressure causes a concurrent rise in speed at a rate of 2 meters per second for every 100 meters of depth. And an increase in salinity causes an increase in speed of 1.3 meters per second for every increase in salinity of one part per thousand.

Thus, electronic analyses of water such as echo sounding must be corrected for changes in the physical characteristics of the water, and the changes in sound transmission can be used to analyze water characteristics.

SUMMARY QUESTIONS

1. What unusual feature of the water-molecule configuration produces its unusual physical characteristics? Describe at least three physical characteristics of the water molecule.

2. Define specific heat and explain how it causes water to possess large amounts of potential energy.

3. What significance may be attributed to the density difference between ice and liquid water?

4. List and describe four important gases found dissolved in ocean water. Explain the biological importance, if any, of each.

5. What is salinity? How is it determined, and how does it vary both vertically and horizontally in the open sea?

6. What is the thermocline? What factors cause it to vary? Give the same information for the halocline.

7. What has been the major source of water contributed to the atmosphere and to the oceans?

8. What is a hydrogen bond, and how does it contribute to the stability of the water molecule?

9. What is meant by latent heat, and how is it important as a property of ocean water masses?

10. What relationship exists between water temperature and the quantity of oxygen dissolved in the water?

11. How do the atmosphere and oceans act together to achieve an equilibrium in composition?

12. What factors affect the salinity of tropical and polar bodies of water? How do the two locales differ?

13. How did the major ocean-water constituents find their way into the sea?

14. How do water temperature and salinity affect the action of sound in water?

15. What are some factors that may affect water color in the sea?

16. What is the halocline?

17. What is juvenile water? How is this related to the formation of the oceans?

7 | OCEAN CIRCULATION

The ocean, like the atmosphere that surrounds the Earth, is dynamic. Large masses of different water types are formed by the mixing of converging water as they move from one latitude to another, and the water bodies on the Earth's surface also move vertically. Vertical circulation forms a vast network of currents which are not yet fully understood.

The total circulation of the oceanic water bodies is highly complex; many factors initiate the circulatory patterns and later modify them.

The primary cause of surface circulation is the pattern of winds that sweep across the face of the Earth in different latitude zones. As the winds move across the water, huge masses of water are pushed along, and large amounts of energy are imparted to them. The general pattern of surface currents is modified by physical factors and the effects of such variables as friction, gravity, the rotation of the Earth, continental formations and their general configuration, topography, and local winds. Each factor interacts with the others to produce complexities in the general flow.

GENERAL SURFACE CIRCULATION

Ocean water is in constant motion; as the water moves across vast expanses of the ocean, it forms huge circular flows that move clockwise in the Northern Hemisphere and counterclockwise in the Southern Hemisphere. As the flow continues, it affects the air masses it encounters and alters the weather and climate of the entire world. Each circular flow, or *gyre*, can be divided into a series of smaller flows with varying characteristics due to the mixing of many water types.

Tidal bore—Bay of Fundy. (NOAA)

These oceanic flows may be thousands of times as powerful as any river. As they move along their prescribed paths, food, gases, and marine organisms become distributed throughout a wide range of ocean depths and world latitudes. In addition, heat becomes distributed more equitably throughout the world's major climatic zones as the surface currents move from regions of high heat loss into the more tropical zones and as water masses overturn (see Fig. 7-1).

Although surface circulation is caused primarily by wind blowing across the surface of the water, density differences also play a role in the total net flow. Salinity and temperature variations affect the movement and character of the numerous water types. Nevertheless, study of the major ocean currents reveals that the wind belts of the world produce the strong, well-defined surface currents that are particularly dominant along the western sides of the ocean basins and currents that are less strong and not as well defined on the eastern boundaries.

Each ocean has its own particular pattern of currents that are produced as masses of water move from one climatic zone to another. However, all oceans share a pattern which is similar on all oceanic surfaces since the factors that initiate the currents and modify them are the same throughout the world.

FACTORS AFFECTING SURFACE FLOW

From Fig. 7-1 it can be seen that the currents along the surface of the oceans form a circular series of whorls, or gyres. Further, the flows that occur beneath the surface follow a pattern similar to surface flow. More detailed examination would reveal that in all cases winds, currents, missiles, and other objects moving across the face of the Earth show a slight deviation from a straight path. Although the factors that alter surface flow are complex mathematical models, we can briefly describe the effects these variables have on surface flow.

The Coriolis Effect

A person standing at any point along the current flow in the Northern Hemisphere and facing in the direction the current moves sees that the current veers off toward the right-hand side. If these observations are repeated in the Southern Hemisphere, the current veers off toward the left-hand side. This motion also forms a series of gyres in the ocean which flow in a counterclockwise direction, rather than clockwise as in the north (Fig. 7-2).

This persistent movement of objects toward the right-hand side of the path observed from the point of origin is called the *Coriolis*

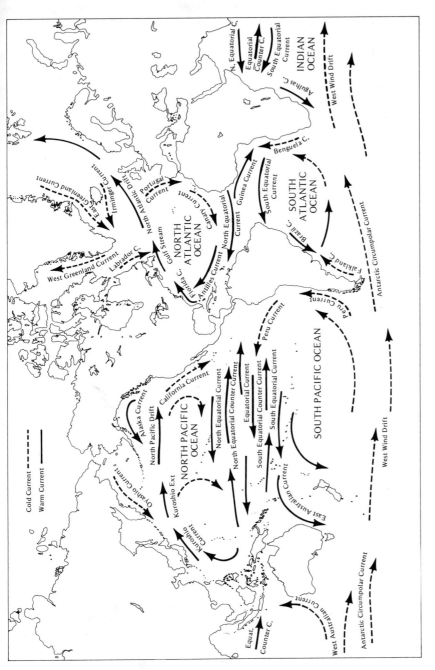

7-1 The wind-driven surface currents of the world.

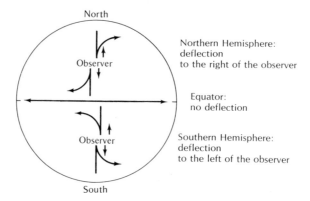

North

Observer

Northern Hemisphere:
deflection
to the right of the observer

Equator:
no deflection

Observer

Southern Hemisphere:
deflection
to the left of the observer

South

7-2   The Coriolis effect at various points on the Earth's surface.

effect, first described by G. G. Coriolis in the nineteenth century. It is an apparent motion imparted to objects moving across the Earth's surface by the Earth's rotation beneath them. The effect is related to a physical property of all moving objects called *angular momentum*.

Angular momentum is expressed in the formula

$$AM = mvr$$

where *m* represents the mass of the object in motion, *v* represents its velocity, and *r* is the radius of the circle (in this case, the latitude of the Earth) over which the object is passing.

This property is used for spectacular feats by the ice skater who goes into a rapid spin as he moves his outspread arms closer to his body and slowly drops to one foot. As he rises from the lowest position and spreads his arms, he slows down again. The mass of his body remains the same throughout the spin, but as he compresses the area over which his body is spread, in order for the left side of the equation (angular momentum) to remain the same, the difference is made up by the increase in velocity of his spinning body. Conversely, as he spreads his arms and raises his body, the velocity decreases and angular momentum remains the same.

The Coriolis effect also influences moving objects as they cross the Earth's surface. As a current of water in the ocean—or a wind in the atmosphere—moves from the equator to the poles, it crosses parallels of latitude which are smaller and smaller. Parallels of latitude are circles which become smaller as one moves north or south; the only parallel that is a great circle (the same as the diameter of the Earth sphere) is the equator. As the current moves across the decreasing radii of the lines of latitude, its relative speed increases *in respect to the rotational velocity of the Earth* and it moves faster than the Earth's rotation. Thus, the current veers off toward the right of the observer. The movement also affects the path of a rocket fired from south to north. Again, the motion is an *apparent* motion, as the speed varies only because the velocity of each successive Earth latitude varies due to decreasing size (see Fig. 7-3).

The same object—current, wind, or rocket—moving from north to south will appear to slow down in respect to the increasing radius of each successive parallel of latitude, and the motion of the moving object lags behind that of the Earth's rotational velocity. Thus, the object will again veer off to the right. The apparent motion due to the Coriolis effect will cause a movement to the left in the Southern Hemisphere, a mirror image of what takes place in the Northern Hemisphere.

In tracing the current flow of each water gyre in each hemisphere, the result of the Coriolis effect can be seen. In the north, no matter where an observer places himself, if he faces in the direction of current flow, the current will tend to veer in the direction of his right-hand side. Wind patterns, which are the primary cause of surface-water flow, show the same pattern (see Chap. 15). Of course, the pattern is affected by other factors; the major modification of gyre motion is caused by the shape of the continents encountered by the moving water.

The Antarctic surface water is the chief water flow uninterrupted by continental landmasses. Here the water circulation forms a gigantic flow from west to east, the West Wind Drift. The main forces affecting the motion of this current are the wind and the Coriolis effect.

7-3 The deflection of a moving object across the Earth's surface; a polar view.

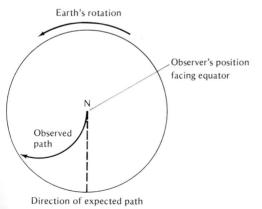

The Ekman Spiral

As wind is the initial and primary driving force for surface currents, the flow of water would be assumed to be parallel to the movement of large-scale winds across the face of the Earth. However, the actual net movement of the water is at a slight angle to the right of the wind, and, further, the current flow along the ocean surface is not effective to all depths but is usually limited to a few hundred feet.

The flow of a mass of water actually occurs as a series of sheetlike movements called *laminar flow.* As each layer of water begins to move, a frictional drag occurs on the water molecules immediately below the surface layer. This drag produces a turbulent flow with irregular eddies. Thus, each successive layer is produced by slippages occurring between successively deeper layers of water. The motion of each layer is slightly to the right of the one above. A series of spiral motions results, rather like a spiral staircase, with each step slightly to the right of the preceding one. Each layer is affected slightly less than the preceding layer, and each successive deflection is a bit more to the right of the one above. The total net diagram has been called the *Ekman spiral,* a model developed to describe the total movement throughout the mass of water (Fig. 7-4).

In the Northern Hemisphere, then, laminar flow means that each mass of water moves to the right of the wind at an angle of 45° along the surface. The net effect is the result of each successive layer's moving more and more to the right of the wind. At some depth below, the water may actually move at an angle of 180° to the initial wind direction. The reversal shift generally occurs at a depth of about 100 meters (300 feet). The total net movement of the entire mass of water is about 90° to the right of the driving wind (see Fig. 7-5).

7-4   Formation of the Ekman spiral.

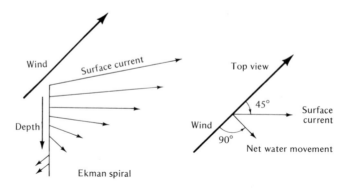

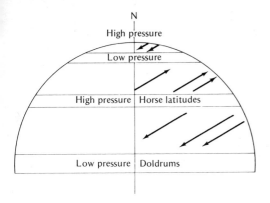

N

High pressure

Low pressure

High pressure | Horse latitudes

Low pressure | Doldrums

7-5 The general circulation of the atmosphere.

THE ATLANTIC OCEAN

In each hemisphere, a great mass of air moves across the ocean's surface in a powerful flow of air called the *Trade Winds*. The Trade Winds in the Northern Hemisphere move from the northeast toward the southwest between the equator and 30° north latitude. In the Southern Hemisphere the pattern of flow comparable to the northern Trades moves from the southeast to the northwest. Each of these wind flows creates the beginning of the oceanic gyre of circulation in the respective hemisphere. The southern Trades create the *South Equatorial Current,* and the northern Trades produce the *North Equatorial Current.* These currents move west parallel to each other on either side of the equator (Fig. 7-6).

The equatorial currents do not coincide exactly with the geographical equator. Instead the water flow is more closely associated with the *meteorological equator,* that is, the zone of the Earth which represents the seasonal shift of the sun's direct rays throughout the year. The meteorological equator varies from about 5° north latitude to about 7° south latitude.

The two equatorial currents are separated by a flow of water which runs toward the east immediately north of the equator. This flow, called the *Equatorial Countercurrent,* is fed by the two equatorial currents and is also a surface flow; it is not a subsurface flow like the countercurrents mentioned later in this discussion. The Equatorial Countercurrent is found in the equatorial region of the world known as the *doldrums,* named from the fact that in the days of sailing vessels, ships could make very little headway here. It is an area of converging air masses from the north and south. The winds in this

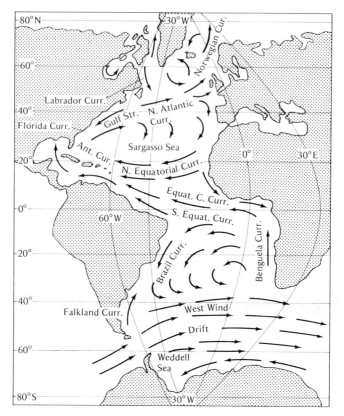

7-6 The surface circulation of the Atlantic Ocean. *(After G. L. Pickard, "Descriptive Physical Oceanography," Pergamon Press, 1963)*

region are light and do not persist. Thus, as the equatorial water returns from the west, it meets very little resistance and is fed primarily by the buildup of water on the western sides of the two equatorial currents.

In the region near the West Indies, the North Equatorial Current separates into several flows. One portion moves into the Caribbean Sea and joins the *Antilles Current.* The rest joins the *Bahamas Current,* that flows past that West Indies island chain. These warm-water masses are in the area that in the hot and humid summer months becomes the primary breeding ground for Atlantic hurricanes. The enormous quantities of latent heat stored in these waters may pass into the air above and produce the extremely low pressure of the tropical cyclones that batter the Gulf Coast and southeastern portion of the United States.

The Antilles Current moves out of the Antilles and joins several masses of water that eventually flow into the *Florida Current* at the Straits of Florida. This water flow is also fed by some of the water from the South Equatorial Current that is forced north as the current is split in two by the point of land at Brazil. The various water masses forced through the Straits of Florida create an enormous and powerful flow through the narrow strait. The buildup of water may reach as much as 7 inches above the southernmost masses pushing through this region. As the water flows north, it joins with the North Equatorial Current through the Yucatan Channel.

All this water meeting in the region of the Caribbean forms an irregular flow through the Gulf of Mexico and joins as an enormous flow past the southeast coast of the United States. This water, known as the *Gulf Stream,* begins to flow north along the coast from Florida. This great warm-water current moves northward until it turns east toward the open sea near Cape Hatteras. There it slows down and begins to mix with cooler water from the north. The current moving across the open Atlantic Ocean, now called the *North Atlantic Drift,* moves toward Norway and eastern Europe, mixing with cooler water from the region around Greenland and the Scandinavian Peninsula. The North Atlantic Drift sends tongues of water north along Scandinavia and past the British Isles which keep the coast of Scandinavia relatively free of ice throughout the year. As these waters come into contact with the air masses near the cold landmass of the British Isles, the famous London fog results as the vapor condenses from the air (see Fig. 7-7).

As the water mixes with cooler water and is itself cooled, it begins to flow southward once more, past the continent of Europe. This southern-flowing current, the *Canaries Current,* eventually reenters the North Equatorial Current.

The warm waters of the Gulf Stream produce the semitropical climate of southern Florida that makes this part of the state a popular vacation spot in winter. Much of the warm water passing the west coast of Spain near the latitude of Madrid originated in the Gulf Stream. Although the Canaries Current is cooler than the Gulf Stream, it is still moderate enough to give Madrid a warmer climate than that found in New York City, at the same latitude, 41° north. New York is most affected by the colder *Labrador Current,* which comes south along the northeast coast of the United States before moving east as part of the North Atlantic Drift.

In the Southern Hemisphere, the South Equatorial Current splits in two at Brazil, the southern part moving along the east coast of Brazil as the *Brazil Current.* This warm-water flow encounters the

7-7   Pollutants mixed with fog induce smog. *(New York City skyline)*

---

north-moving, cold *Falkland Current* along the southeastern coast of South America. Both currents then mix and move toward the open sea and flow east toward Africa. The cooler current thus formed begins to flow north once more along the west coast of Africa. This current, the *Benguela Current,* brings relatively cool conditions and frequent fogs along the western coast of South Africa. The Benguela Current finally joins up with the South Equatorial Current and completes the gyre in the North Atlantic Ocean.

THE PACIFIC OCEAN

In the Pacific Ocean, the North Equatorial Current flows from east to west. A portion of this current flows northward toward Japan and becomes the *Japan Current,* or *Kuroshio.* When it encounters cold water from the north, called the *Oyashio,* it mixes with it. The water mass, cooled by the Arctic water, flows toward the American continent as the *North Pacific Current* and flows south along the coast as the cool *California Current,* continuing south until it rejoins the North Equatorial Current off the coast of Baja California.

In the South Pacific Ocean, the South Equatorial Current enters the *West Wind Drift,* or *Antarctic Current,* in the southwestern part of the ocean. The current then flows north along the west coast of South America as the cool *Peru,* or *Humboldt, Current.* This current was noted by Alexander von Humboldt as the cause of the cool, foggy conditions which prevail along this coastline. The waters then reenter the South Equatorial Current, completing the South Pacific gyre as a counterclockwise mirror image of the North Pacific gyre (Fig. 7-8).

THE INDIAN OCEAN

The Indian Ocean has many variable currents due to the monsoon winds that blow across it, changing direction as the seasons change. In the western half of the Indian Ocean water is exchanged with the Red Sea, bringing dense, saline water into the Indian Ocean.

There is one small gyre of water in the northern part of the Indian Ocean and another in the southern half. The northern gyre does not persist and does not give rise to any currents of great importance, but the southern gyre gives rise to the *Agulhas Stream,* a strong current that moves along the southeast coast of Africa (Fig. 7-9).

THE POLAR REGIONS

In the Arctic a gyre flows between Alaska and Japan. From the south coast of Alaska the *Alaska Current* flows toward Japan and Siberia; it moves south and gives rise to the current called Oyashio, mentioned earlier. The current then returns to the *Subarctic Current* immediately north of the North Pacific Current and flows in the same direction toward North America.

7-8   The surface circulation of the Pacific Ocean. *(After Pickard)*

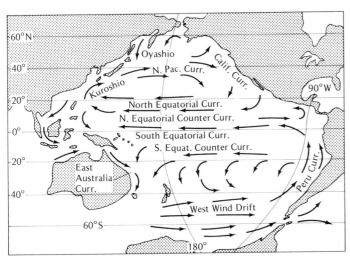

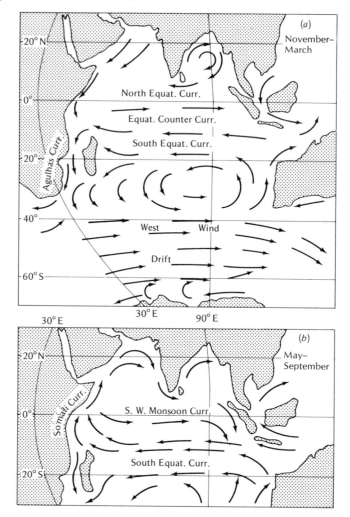

7-9   The surface circulation of the Indian Ocean. *(After Pickard)*

The Antarctic Ocean has the West Wind Drift flowing around the southernmost continent. Fingers of water flow north along the southeast coast of South America as the Falkland Current, along the southwest coast of Africa as the Benguela Current, and along the east coast of Australia as the *West Australia Current.*

DEEP-WATER MOVEMENT

The most important consideration in investigating the general movement of water masses throughout the oceans is density. Density differences between adjacent masses of water cause vertical movement of water and create the deep-water masses that move through the oceans in slow, sometimes undulating systems. These masses have their eventual influence even on surface flow. The density differences between water masses are due primarily to variations in their salinity and temperature. The density effects create a *thermohaline* circulation, a form of movement directly influenced by density differences resulting from different combinations of salinity and temperature. Although other factors affect deep-water circulation, thermohaline effects are of primary importance.

Of course, because of the great depths at which this circulation occurs, much information is lacking about bottom-water circulation. In addition, the bottom-water circulation is affected by bottom topography, about which geological information has been lacking until quite recently.

THE LAYERED OCEAN
Antarctic water is of primary importance to studies of bottom circulation because this dense, cold water moves from the Weddell Sea and sinks to the bottom of the ocean. The water spreads along the bottom and reaches into the northern seas. In the North Atlantic Ocean, in particular, this water meets no resistance from bottom waters. Rocky sills separate the Arctic from the Atlantic Ocean and only a small, less saline flow of bottom water enters the North Atlantic from the Arctic. Thus this smaller flow of northern water cannot interfere with the larger flow of water from the south.

Surface temperatures taken in the southern oceans reveal a slight surface-temperature increase which continues from the equator to about 50 to 60° south latitude, where a sharp increase appears at about 35 to 40° south latitude. These temperature increases represent regions where, at the first increase, Antarctic water masses converge and at the second subtropical water converges. Cold, dense water sinks in each zone and mixes with the bottom waters in each region (Fig. 7-10).

Thus, deep-water circulation is slow and primarily under the influence of density differences between adjacent water masses. Little is understood about the complete movement of the water masses below the surface.

In general, the pattern begins as cold water at either pole. The density of this water becomes greater as ice forms, concentrating salt in the remaining water. The cold polar waters sink and slowly move toward the equatorial regions. The flow from the Antarctic is more pronounced as the Antarctic bottom waters are found in the Northern Hemisphere as cold bottom water.

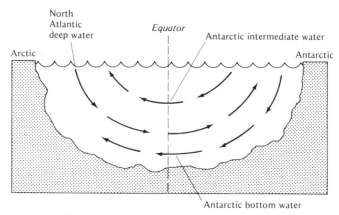

7-10   A generalized view of the subsurface flow of the ocean basins.

Above the Antarctic bottom water is a second flow, also produced in the Antarctic. This water is less dense and flows above the Antarctic bottom water as the Antarctic intermediate water.

The two Antarctic water flows are separated by a southward moving water mass. Called the North Atlantic deep water, it is formed primarily as sub-Arctic water near Greenland. Most of the potential Arctic flow is trapped in the north by rock sills around the South Arctic Ocean in the Western Hemisphere.

All these water movements are influenced by the Coriolis effect, but they follow a general north-to-south or south-to-north movement. Bottom topography also influences their flow, as they encounter ridges which interfere with free circulation.

COUNTERCURRENTS

The mighty Gulf Stream contains nearly 50 times more water than all the world's rivers combined. It also gives rise to several minor currents off the Grand Banks of Newfoundland. But this current and all the others in the world are much more complex than they first appear. Beneath these surface currents deep-water countercurrents flow. Although the details have not been completely worked out, we know that there is a vast system of countercurrents which may be as pervasive and important as the surface currents already charted by oceanographers.

For example, below the westward-flowing South Equatorial Current flows the *Cromwell Current,* first discovered in 1952. It moves east at a depth of about 300 feet and is at least 4,000 miles long. The water in this current flows at a rate of 30 million tons per second. Added to this eastern flow is the surface countercurrent. Together, these two flows of water carry 60 million tons of water from west to east.

Beneath the Gulf Stream a counterclockwise flow of water arises from deep, cool water south of Greenland. This cold, dense water begins as surface water and sinks toward the bottom. It feeds into the North Atlantic Ocean at a depth of 1 to 2 miles, and traveling beneath the Gulf Stream flows into the Southern Hemisphere and feeds a current running beneath the weaker Brazil Current and in the same direction.

GENERAL BOTTOM FLOW OF WATER

Bottom water in all the world's oceans generally originates in the polar seas. This cold water becomes highly saline as ice forms, leaving salt in the water in greater and greater concentrations. The waters, in general, sink to levels which are in equilibrium with their varying densities; as they sink, they move toward the equator. The polar water is replaced by less dense intermediate water that forms in the central basins and flows toward the poles.

Thus, the general flow of Antarctic bottom water tends toward the north. This high-density water sinks as Antarctic and sub-Antarctic water and spreads outward as it moves north. The flow is affected by bottom topography such as the ridge-rift system and is modified as the water mixes with other water masses. Some of the water returns to the Antarctic as it mixes with south-flowing surface and intermediate water.

North Atlantic deep water results from a mixture of high-saline Gulf water and the less saline Arctic water near Greenland. This sinks toward the bottom and moves south toward Antarctica. The North Atlantic deep water and Antarctic bottom water meet at about $20°$ north. Here, the saline North Atlantic water flows above the Antarctic bottom water.

The general movement of bottom waters forms a layered effect in all the ocean basins. The motion is rather slow. On the average Antarctic bottom water may take several decades to reach the equator, and some bottom flows may require hundreds of years to complete a single cycle. But the motion of bottom waters does influence water throughout all depths and has an effect on the upper levels of water which interact with the atmosphere. The exact nature and configuration of bottom circulation, subsurface countercurrents, and their allied influences on all levels have yet to be fully worked out.

SUMMARY QUESTIONS

1. What two factors affect the primary horizontal and vertical circulation of the ocean? Describe the Coriolis effect and its influence on circulation.

2. What major factors are responsible for the general flow of surface circulation? What additional factors affect subsurface flow?

3. In general, what is the direction of flow of warm currents and what part of the continents do they primarily affect in each hemisphere? Describe the general flow and results of cold currents in each hemisphere.

4. What factors influence the flow of Antarctic water, and what water masses are created by the Antarctic surface water?

5. What are countercurrents? Describe one.

6. Why may the ocean be described as a layered ocean?

7. What is angular momentum? How is it expressed?

8. How do parallels of latitude relate to changes causing the Coriolis effect?

9. What is the Ekman spiral? What does this model describe?

10. Describe the general flow of water in the Atlantic Ocean in the Northern Hemisphere. How is this flow separated from the Southern Hemisphere?

11. How does the Gulf Stream affect Florida? Why does the southern half of the eastern seaboard of the United States differ in climate from the northern half?

12. What is meant by thermohaline circulation? How is it related to surface and bottom flow?

13. Describe the general flow of bottom water in the polar regions and through each ocean basin.

14. What is meant by laminar flow, and how does it occur in the sea?

15. What leads us to suspect that there may be a vast undersea flow running in the opposite direction to the major surface currents? Explain your answer.

16. Why is the Indian Ocean surface less clearly defined than other oceans?

17. Describe angular momentum as it affects moving objects.

18. How does the meteorological equator differ from the geographic equator?

19. At what latitudes does ocean surface temperature show an increase? Why?

20. At what latitude does Coriolis have no effect?

21. Describe the general surface flow of atmospheric circulation.

8

TIDES AND OTHER MOTIONS OF THE SEA

We have seen that the waters of the oceans are in a state of constant flux. Surface currents carry warm water from the low latitudes and distribute the energy into the colder climates. Conversely, the cold waters of the polar regions move into the warmer regions of the world, where they eventually mix with different water types. At the same time, the waters at depth circulate in a variety of ways quite different from circulation at the surface.

Winds are the primary driving force behind the surface currents, but, in general, bottom currents move as the result of density differences due to thermohaline effects. These bottom waters move more slowly, but their circulatory patterns are even more complex than those of waters which move across the face of the ocean. In addition, many factors modify the circulation of both surface and bottom water.

However, these movements are primarily the result of wind and density and do not take into account other motions of the sea that are often just as regular and pulsating as the general circulatory patterns. Among these other motions are tidal effects and wave motion, both of which act against the shorelines of our continents.

Tides are the result of regular motions of the Earth and moon as they revolve about a common center of gravity between them. Waves are produced by wind effects on surface waters as they approach the coast and are altered by the rise of the continental shelf and the configuration of coasts. As these energy waves pass through the water, they create their own typical patterns of water movement and change the shape and configuration of the shoreline.

Tide staff. *(NOAA)*

It must be mentioned at the outset that one well-known feature of oceanic movement is not due to tidal forces or normal wind-driven wave action. *Tidal waves* that intermittently wash over the coastline in various parts of the world are not due to normal tidal forces. These waves, which range in size from less than 1 meter (3 feet) in height to well over 30 meters (100 feet) are the result of earthquakes in the floor of the ocean. They are properly termed *seismic sea waves* or *tsunami,* a name taken from the Japanese. These energy waves of gigantic proportions moving through the water result from the enormous quantities of energy imparted to ocean-water masses as the earthquake vibrations pass into the water from oceanic seismic activity. As the surge of water approaches the coastline of a landmass, the wave builds in much the same way as a wind-driven wave and wreaks its havoc on the coast when the energy of the water is expended upon the lands.

Several other kinds of movement require brief discussion, for example, internal waves, turbidity currents, and rotary currents. But the chief motions we are concerned with are the two energy surges that regularly pass through the oceans and affect our coastlines, tides and waves.

TIDES AND WAVES

Of all the circulatory patterns that occur in the ocean, wave action and the daily movements of the tides are best known to the average person. As one stands on the shoreline it is relatively easy to observe the changes in the tide by reference to debris and water marks left at high tide. Waves of various heights can be noted, and although their relation to topographic effects may not be obvious, observations made during calm periods and storms certainly reveal the effects of wind on wave characteristics.

TIDES

Tides are the direct result on the body of water covering the Earth of gravitational effects produced by the sun and moon. These daily, regularly occurring movements are further accentuated by an oscillating harmonic motion imparted to the water by the shape of the basin containing it. As the vibrations pass through the water, a regular pulse of harmonic motion passes through the water as a result of the configuration of the basin. This harmonic effect may be compared to the regular pattern of oscillations occurring in a bowl of water being carried from one place to another. The water washes back and forth across the basin due to the energy train that moves through it. The shape of the basin also influences the motion of the water.

Whereas waves generally expend their enormous stores of energy scouring the coastlines and topographic features against which they break, tidal effects are not limited to this activity. It can be shown that tidal energy and motion has affected the rotational velocity of the Earth. Friction between the Earth and the tidal bulge moving around the planet slows the Earth's rotation on its axis by an almost imperceptible amount each year. It is estimated that the rotation of the Earth slows by approximately $\frac{1}{1,000}$ second each century. Added to this is the assumption that the moon apparently was once closer to the Earth than it is at present. If true, this would mean that the effect of tidal friction was more pronounced than it is now.

Although this drag seems slight, the cumulative effect of this slowing becomes significant over the vast stretches of geologic time. We assume that at least 5 billion years have passed since our Earth-moon system was born.

Marine biologists have analyzed the growth of certain corals in order to ascertain changes in our rotation. These creatures exhibit a daily pattern of growth rings similar to the annual growth rings in trees (see Chap. 5). Analysis of the daily growth cycles in these corals reveals that the yearly revolution of the Earth around the sun during the Devonian period (about 400 million years ago) was at least 400 days long and the days were much shorter than at present. The frictional drag between the tidal bulge of the ocean waters and the Earth rotating beneath has effectively slowed the rotation of the Earth. As the Earth's rotational velocity slowed, the length of the day increased and fewer days occurred during each yearly revolution around the sun.

What produces this enormous effect on the waters of the Earth? How can such a fundamental change in the motion of the Earth be caused by what seems to be merely a slight change in the height of the water along the shore?

The tides are explained initially by Newton's universal law of gravitation. This law, which explains how our solar system is kept in balance and which established modern astronomy on a firm foundation, states that any two objects attract each other with a force that is directly proportional to their mass and inversely proportional to the square of the distance between them.

Essentially, this law tells us that as the Earth causes the moon to revolve around the Earth, the moon also exerts an effective pull on the mass of the Earth. Actually, the Earth and moon act as a single system, and it is erroneous to think of the moon as merely revolving around the Earth because the Earth and moon revolve around a *barycenter,* a common center of gravity for both masses. It is rather like the two knobs at the ends of a twirler's baton spinning about some fulcrum that is not at the exact center of the baton. In the Earth-moon system, the barycenter lies about 2,900 miles from the center of the Earth.

As the moon moves around the barycenter, its gravitational force causes the waters of the Earth to bulge outward on the side of the Earth facing the moon. At the opposite side of the Earth, centrifugal forces produced by the Earth's rotation cause a second, slightly lower bulge of water to be thrown outward. Thus, the Earth has two bulges, or high tides, which appear in a straight line with the moon and two troughs, or low tides, which appear between the highs at right angles to the moon. As the moon moves around the barycenter, the bulge of water follows beneath it, bringing all parts of the Earth into contact with the highs and lows during alternate periods (Fig. 8-1).

The position of the tides shifts slightly during the year because the moon does not follow the same path around the Earth every day.

8-1   The forces producing tides in the Earth's bodies of water.

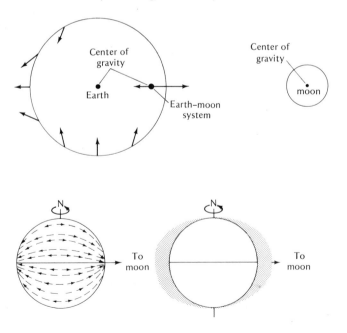

During the passage of a single year, the moon shifts its position from 28.5° north latitute to 28.5° south latitude. This change in the moon's position over the Earth causes the bulge to shift with the moon. Thus, the tides in any single locale are not the same throughout the year. The tidal heights vary somewhat depending upon the exact position of the moon and the exact position and configuration of the tidal bulge.

The tides do not pass around the Earth at precisely the same time each day throughout the year, a characteristic that is also due to shifts in the moon's passage around the Earth. The moon passes over each locality about 50 minutes later each successive day. Thus, the tides occur 50 minutes later than the day before. If the moon passed overhead at precisely the same time each day, we would have no need for tide tables, as tides would occur at the same time each day; this would be especially true if the moon revolved parallel to the equator of the Earth.

In addition, the tides do not coincide exactly with the moon's position as it passes each given point on the Earth's surface. Due to the friction between the Earth and the tidal bulge, the tide occurs, on the average, 59 minutes after the moon's passage. However, this factor is apparent only in the delay in the daily tide.

One other major factor influences the height of the tidal bulge, the gravitational zone of the sun. Since the sun is a massive object, its effects can be noted on the tides although they are only about one-half those of the much smaller but closer mass of the moon.

When the sun, moon, and Earth are in a relatively straight line in the heavens, we see the *full moon* and the *new moon*. At these two periods, the effect of the sun is added to that of the moon and we experience the widest range in tides. That is, the differences in height between high tide and low tide are greatest. This situation is known as *spring tide*. When the moon, Earth, and sun are at right angles to one another, the moon is said to be at *quadrature*. This situation produces the smallest ranges in tide and is referred to as *neap tide* (see Fig. 8-2).

Thus, a variety of factors affect the ranges between tides, the tidal height, and time of occurrence.

Tides and Basins

The actions of tides also depend upon the shape of the basins holding the water. The shape and configuration of the basins influence tidal motion as water moves in and out. The freedom of the movement of water in semienclosed or open bodies also modifies tidal movement. The shape and type of water body superimposes various mo-

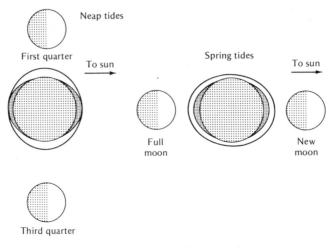

8-2   The formation of spring and neap tides.

tions on the regular rise and fall of the water. These motions are a series of oscillations which vary according to the size and configuration of the basins.

As the tide moves across the face of the Earth, tidal energy creates a series of waves in the ocean. This progressive series of waves passes across the surface and is altered by the shapes and placement of continental landmasses and local conditions determined by the shape and configuration of the coastline.

For example, in some small basins and bays, the tidal wave acts simultaneously throughout the basin with all parts of the basin experiencing high and low tide at the same time. In other basins, particularly larger open basins, the tide progresses from one side of the basin to another. The tide alternates with a series of progressive low-water periods.

In addition, in any given period of time, three different types of tides, *diurnal, semidiurnal,* and *mixed,* may be observed on the Earth in different locales. Each type is due to alterations resulting from the shape of the basin in which the tide is occurring, the Coriolis effect, and the local topography.

Diurnal tides are daily changes producing one high and one low tide with progressive changes in water level causing the extremes in range to occur about 12 hours apart. The Gulf of Mexico and parts of southeast Asia experience these changes, in which partially enclosed basins affect the range. For example, the east coast of Korea, which is open to the Pacific Ocean, experiences a tidal range of about 1 foot each day. However, as water moves into the South China Sea on the west, the Coriolis effect and the nature of the basin cause the water to veer to the east and pile up against the west coast of Korea, creating a tide of approximately 12 feet at Inchon, while the South China coast experiences a tide of only 3 feet. Certain inlets in the Bay of Fundy may experience tides in excess of 40 feet (Fig. 8-3).

Semidiurnal tides occur twice each day as on both sides of the At-

8-3   The sequence of tidal ranges in the Bay of Fundy.

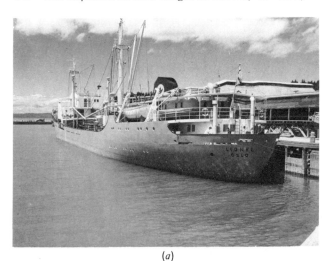

(*a*)

(*b*)

lantic Ocean, on the East Coast of the United States, and on the west coast of Europe. Two high tides and two low tides occur, with 6 hours between each. The ranges are about equal, usually only about 2 to 3 feet on the average.

In *mixed tides,* common on the West Coast of the United States, diurnal and semidiurnal tides are superimposed on each other. In this situation, the ranges occur in the same manner as semidiurnal tides but the ranges between each set are of different heights.

Other Related Tidal Events

As the tides move across the Earth's surface, other motions are exerted upon the water as a result of the tidal forces. *Tidal currents* may be induced as a horizontal flow of water along the coastline and through enclosed areas such as rivers and bays as the water rushes into the channel. On open water, *rotary currents* are induced as a regularly occurring oscillation with each tidal cycle. These currents are obstructed along coastlines and occur as regular currents that flow in toward the coast and reverse themselves during each tidal cycle. In this case, they are termed *reversing tidal currents.*

Internal waves are also noted within water masses. They are not surface phenomena but seem to be the result of density differences in the water body producing a repetitious shallowing and deepening effect in the water mass. These waves are little understood at present but are apparently related to tidal motion.

WAVES

Waves are regularly occurring movements in the water as the result of interactions between the ocean surface and winds that blow across the water (Fig. 8-4). The energy imparted to the water produces wave motion, but the only progressive movement is of the energy through the water, not the water itself. Wave motion is modified by the general contour of the ocean basin, especially as waves approach the shoreline.

Wave size depends upon several factors. The wind speed determines the amount of energy imparted to the water, and the *fetch,* or distance over which the wind blows, determines the amount of energy that can be imparted to the water. Wave size also depends on the duration of the wind. The size of the wave may vary from ripples, which are less than 1 centimeter in height, to gigantic waves that may be in excess of 30 meters (100 feet). In general, *sea waves* are waves directly produced by wind stress on the ocean surface. As these travel outward, faster-moving waves overcome the slower-moving waves and create a regular pattern of waves called *swell.*

8-4   Waves on the open sea. *(NOAA)*

As we shall see in Chap. 9, waves are the most important cause of alterations and evolution of our coastlines. As the waves approach the shore, they erode and transport material; the manner of transport and subsequent deposition are uniquely determined by the shape of the land and the bottom topography over which the incoming wave passes (Fig. 8-5).

Wave action depends on many things, among them wave speed.

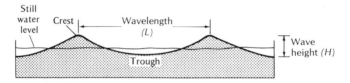

8-5 Wave characteristics.

The speed of an approaching energy train is a function chiefly of the wavelength L and time T according to the formula

$$\text{Speed} = \frac{L}{T}$$

As waves approach the shoreline, the energy contained by the mass of water is expended on the shoreline and its attendant features. The effects of erosion are clear to see in places where waves are unimpeded by protective measures.

As the energy wave enters the surf zone, the wavetrain may "break" several times (instead of just once, as is commonly believed to be the case) before dissipating its energy. How the wave breaks depends entirely upon the nature of the bottom. After the water reaches the shoreline and dissipates its energy, the water returns seaward as a flow called the *rip current*. Water flows parallel to the shore as a *longshore current,* transporting material along the shore and generally leaving coarse materials inshore with finer, more easily transported materials grading outward (Fig. 8-6).

The net transport of material along a beach on the coastline explains the transitory nature of beaches and some islands. Frequently, the continuous action of waves and longshore currents removes sand from one region of a coast or island to another part of the coast. Over extended periods of time, islands and beaches can be observed to migrate in a specific direction.

BOTTOM CURRENTS

Many other movements of the ocean's waters result from local and specialized processes which occur at depth. Among the most significant are those produced during upwelling and turbidity currents.

TURBIDITY CURRENTS

Many areas of the ocean floor and continental shelves are found to be carved out into deep, elongated valleys. In addition, large parts

Spilling breaker

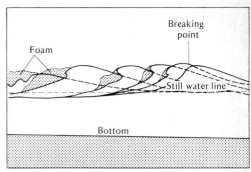

Spilling breakers

Plunging breaker

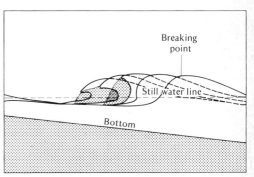

Plunging breakers

8-6 Breaker formation in shallow water.

of the shelves and ocean floor are devoid of sediment. Where does this sediment go? What forces gouged out the valleys found throughout the world? Although there is still some controversy about these points, one cause may be the action of turbidity currents acting on the shelves and floor of the ocean.

Turbidity currents are dense, sediment-laden currents of water which periodically result from the buildup and instability of sediment deposits. As these undersea avalanches occur, they carry large quantities of sediment with them. The sediment acts as an abrasive and scours out the floor and shelves beneath the swift-moving and dense turbidity current (Fig. 8-7).

It is further suspected that turbidity action occurs on the sea floor and has been the chief agent forming the canyons on continental shelves, like those at the mouth of the Hudson, Congo, Amazon,

8-7   A sandfall in the Cape Lucan submarine canyon, Baja California. The height of the fall is 30 feet. *(Scripps Institution of Oceanography)*

---

and other major land rivers (Fig. 8-8). It should be noted, however, that these canyons may be the result of activity other than turbidity currents.

UPWELLING

Upwelling, which was referred to earlier, is common along the west coasts of most continents, although it occurs in many other parts of the world as well (Fig. 8-9).

Upwelling is a direct result of wind stress on the sea surface. As winds blow parallel to a coast, the net flow of water is at an angle of about 45° toward the sea. This flow causes cold bottom water to move upward to replace the warmer surface water moving offshore.

The cold water is rich in dissolved nutrients and can carry a large

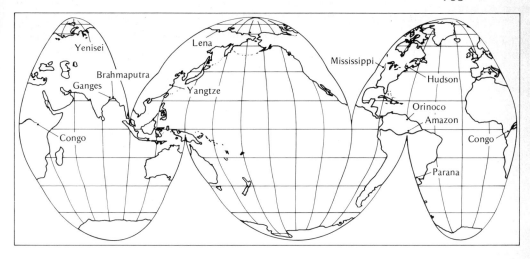

8-8 The distribution of canyons on the continental margins.

supply of dissolved oxygen as it is in contact with the atmosphere. Thus, regions of upwelling are particularly fertile grounds for the support of large fish populations.

However, since upwelling is directly related to a meteorological phenomenon, it can be altered by a change in atmospheric conditions, as demonstrated along the west coast of South America. Periodically, the winds that initiate the upwelling along the coasts

8-9 The regions of upwelling throughout the world.

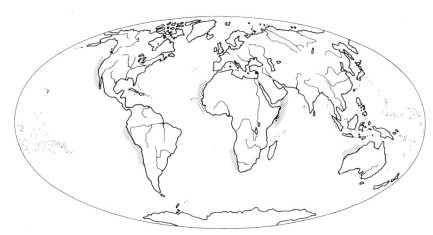

of Peru and Chile drop and fail to drive off the warmer surface waters. Then the warmer waters to the north move south and invade the area. Because this warmer condition occurs most frequently around Christmas time, it has been named *el niño,* "the child." The warmer water cannot support the abundant fish population, and they die by the millions. In addition, a large bird population normally feeds on the fish, and these too die off for lack of food.

The condition reverts to the original upwelling circulation with the seasonal production of the south winds. But the overall effect of the death of the fish and birds as well as the hydrogen sulfide gas in the air is spectacular and frightening, especially when the planktonic *Gymnodinium brevis* colors the water red with its poisonous waste products. This so-called *red tide* is a frequent visitor to other parts of the world as well.

ESTUARIES AND FJORDS

In many regions of the world basins, bays, and inland arms of the sea that are partially enclosed strongly affect the circulation and physical characteristics of the water in the area. The movement of the water is determined in general by factors we have already discussed, but special water motion results from the unusual local topography.

Estuaries, for example, are regions in which a partially enclosed basin expels water into the sea. Many of these on the East Coast of the United States resemble Scandinavian fjords (Fig. 8-10) in being drowned glacial valleys. Others are deep river channels entering the sea. In the mouth of the river flowing through such a region, fresh-water and salt water meet and form a salt wedge at the bottom with a wedge of fresh, less dense water at the surface. The freshwater runs seaward, and the salt water extends toward the land. At times sea-water can be found far upstream beneath the fresher water. The brackish water produced by the intermingling of the two water masses calls for a rather specialized adaptation in the organisms that live in this environment. The salt water may penetrate farther upstream during periods of high tide or when drought conditions reduce the flow of water in the river.

Tidal *bores* are created in certain estuaries during high tide, or flood tide as it is sometimes called. The narrow confining walls of these drowned glacial valleys amplify the effects of the tide-produced wave. The bore thus induced may travel upstream as fast as 15 kilometers per hour and may be 3 meters or more in height.

Further, the results of the Coriolis effect can also be seen in the movement of the masses of water in the estuary. In the Northern

8-10 A fjord in western Norway. *(Norwegian Embassy Information Service)*

Hemisphere, the freshwater is forced to flow to the right side of the channel as it flows seaward while the salt wedge piles up along the left-hand side.

Other areas of the seas show peculiar circulatory patterns as the result of unusual bottom topography. Underwater ridges may interfere with circulation and create basins of differing properties even within a single large inland sea.

For example, the Black Sea has only a small outlet through the Bosporus. In this region, a high rate of runoff from land rivers and much precipitation bring large amounts of freshwater into the Black Sea. The surface water is low in density and carries a great deal of oxygen. The surface waters of the sea abound in fish, but the rocky sill separating the Black Sea and the Bosporus slows the circulation of bottom water. Below about 200 meters (600 feet) there are large concentrations of hydrogen sulfide gas, and the sea bottom is almost totally devoid of life.

Some fjords in Scandinavia have a similar configuration. Seaward sills of rock cut the bottom waters off from the sea. The upper waters are fertile fish grounds, but the bottom waters, dead and stagnant, contain little oxygen.

The floor of the Mediterranean exhibits a geology exactly opposite that of the Black Sea (Fig. 8-11). A sea-floor ridge between Sicily and North Africa separates the Mediterranean into two distinct basins. The shores of the eastern basin are an arid region, where little freshwater is fed into the sea. In addition, a high rate of evaporation causes the sea to have an above average level of salinity. The salinity approaches 39 parts per thousand in the eastern basin.

The western basin has a small outlet to the Atlantic Ocean past the Straits of Gibraltar. The dense water slowly moves from the eastern basin across the central sill and sinks to the bottom in the western basin. This water continues west and moves across the ridge at Gibraltar into the Atlantic Ocean as a bottom current. Less saline Atlantic Ocean water enters the Mediterranean Sea across the top of

8-11   A cross section of the Mediterranean basin.

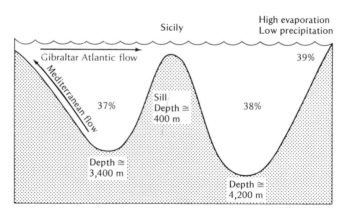

the dense, saline Mediterranean water. The dense, saline Mediterranean water can still be identified as a single mass of water hundreds of miles to the west in the Atlantic.

During World War II, this unusual pattern of dense water moving west and upper-level Atlantic water moving east was exploited by German submarines to slip past the British blockade of Gibraltar. The submarines would turn off their engines and let the Atlantic Ocean water push them into the Mediterranean. To leave the sea, they would drop into the denser water along the bottom of the western basin and move out with the current flowing toward the Atlantic.

SUMMARY QUESTIONS

1. Why are tidal waves a misnomer? Explain how tides are produced and what is meant by the daily tidal range.

2. How have tides affected the length of the day? What evidence have we for this phenomenon?

3. What are the differences between neap and spring tides?

4. How are surface waves produced? Explain wave action on the open sea.

5. What are turbidity currents? What effects do we suspect them to have on the shelves and ocean floor?

6. What produces upwelling? What important economic activity depends on it?

7. Briefly describe an estuary and how water circulates in such a region.

8. Explain the relationship between tides and the sun and moon. How do these two bodies affect the tides?

9. Who was the investigator who described the effect of tides in relation to the Earth-moon system? What discovery led from the statement of tidal forces?

10. Why is there a variation between the tides and the passage of the moon over each place? Why do the tides occur at a different time each day?

11. Describe the three types of tides.

12. How may the shape and configuration of the basin affect the type and movement of tides?

13. What is a tidal bore? How is it related to tides?

14. How does the unusual configuration of the Mediterranean Sea affect its circulation?

15. What effect might the tidal and wave energy have on an island over an extended period of time? Will the action be persistent and consistent?

16. Why is upwelling a major factor in producing rich fishing grounds?

17. What is a barycenter? Where is it for the Earth-moon system?

18. Why is there a great contrast in tides between the east and west coasts of Korea?

19. What are internal waves? Of what are they the result?

20. What is sea swell?

21. How is wave speed calculated?

22. What is a turbidity current?

23. What features are attributed to turbidity currents?

24. What is the red tide? How is it produced?

25. Describe the unusual difference between the surface and bottom waters of the Black Sea.

26. What event causes upwelling to cease?

27. What is a long-shore current? What effect does it have on the shore?

9 | WAVE ACTION

The surface of the ocean is set in motion by the action of winds blowing across it. As a result of the transfer of energy transmitted to the water by the wind, waves are formed on the sea surface. Waves are an easily identified feature of the sea. Although winds usually initiate surface waves, waves also may be generated by the gravitational attraction between the Earth and the moon and the sun. In rarer cases, waves also may be produced by the dynamic forces which disturb the Earth's crust and produce earthquakes.

The study of waves moving onto the land probably represents one of man's earliest attempts at studying the movements of the ocean. Yet, in spite of the age-old study of the movement of water across the surface of the sea, our understanding of the action of the waves is still in the embryonic stage.

Waves continue to beat unceasingly on the shores and coastlines of the continents, moving up in a graceful fashion only to crash onto the continent as the wave breaks. The waves act to alter the coastline as they pound the exposed shores. In times of storm activity, the waves increase in magnitude and velocity to produce widespread damage and wreak havoc on the lands bordering the oceans.

For a better understanding of the predominant force producing and altering the beaches and coastline features of the world we need a description of the physical features of waves. The characteristics of waves can be explained using an idealized waveform to describe wavelength, wave height, and wave period (Fig. 8-5).

Wavelength is the horizontal distance separating points in a similar position of oscillation on successive waves. For example, the distance from one wave crest to the next wave crest represents the wavelength. The letter L is used to represent the wavelength.

The wave height H is the vertical distance of the wave measured from the wave through to the wave crest. The wave height depends on the energy of the wind and is independent of the wavelength. However, the height of a wave cannot exceed one-seventh of its length without becoming unstable and breaking. The ratio of H/L describes the steepness of a wave and is useful in explaining the effect of wind velocity on the sea surface.

The time required for successive waves to pass a given point is the period of a wave. Since period is a factor of time, the letter T is used to represent the wave period, which is expressed in seconds.

As we have indicated, the height of a wave and the period of a wave are related to the wind. In addition to the velocity and time over which the wind acts, the water-surface distance over which the wind blows is important. This distance is referred to as *fetch*. In general, fetch affects the length and height of a wave. The greater the fetch the longer the wavelength and wave period. The relationship between a short fetch and wave height is direct; that is, for waves with short fetches the wave height varies directly with the wind velocity. The same relationship for a long fetch is inverse; that is, it tends to produce a short wave height. Therefore, waves having a long period can only be generated by a long fetch.

Since waves depend on all these mentioned factors, any change in one of them will produce waves in which physical characteristics vary considerably. It is apparent that the waves produced in oceans and lakes will be quite different (see Fig. 9-1). The small ripplelike waves produced on a small lake are certainly quite different from mighty waves pounding the coastline.

In order to understand the motion of water fully it is necessary to separate one wave from the group and consider it as an individual. This, of course, is not the way waves behave, but for the purpose of our explanation it will serve as a model. The same forces acting on a

9-1 A typical waveform.

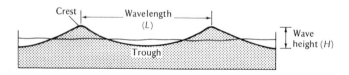

solitary wave through the energy-transfer process will approximate the energy of each individual wave in a wavetrain. The motion of a molecule of water in a wave is quite different from the apparent motion of the wave itself. A distinction must be made between the movement of the water particles making up the wave and the movement of the waveform.

PARTICLE MOTION IN WAVES

If the motion of an object floating on the ocean surface is observed carefully, it is seen that the object does not travel with the wave but bobs up and down as the wave passes. Closer examination shows that the bobbing object does not simply go up and down but appears to move in a circle (see Fig. 9-2). As each successive wave passes a point, the object makes one complete revolution.

The floating object may be represented as a particle of water at the wave crest. The same observation shows that the particle at the crest moves forward with the wave crest in the direction of the wave ad-

---

9-2 Particle motion in a wave entering shallow water.

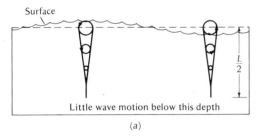

Surface

Little wave motion below this depth

(a)

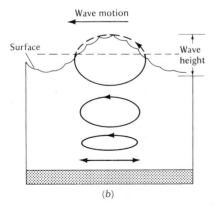

Wave motion

Surface

Wave height

(b)

vance. As the particle approaches the wave trough, it moves backward, opposite the direction of wave advance. Between the crest and the trough the particles all seem to move up and down. Thus, the water particles oscillate in a circular path, returning to the original position after one complete cycle.

The particles tend to move slightly faster in the wave crests than in the wave trough, resulting in a slight horizontal displacement in the direction of wave travel, but since the wave velocity is so much faster than the particle advance, the net advance is negligible.

The velocity of the water particles decreases with depth. The particle motion is at a maximum at the surface but is greatly reduced with depth. The radius of the circular motion is reduced and becomes more elliptical with depth until the particle no longer moves in a circle but back and forth horizontally. The orbital motion decreases below the surface to a depth equal to one-half a wavelength, at which point it ceases to be apparent. Wave-generated circular water movement is clearly a surface phenomenon in the deep ocean.

Whenever the depth of the water is less than one-half the wavelength, the bottom interferes with water motion as the wave begins to "feel" the solid bottom. As a result, the wave height increases and the wave steepens to the point where it breaks. This generally occurs when the particles at the crest of the wave move faster than the advancing waveform, causing the wave to become unstable.

Waves approaching the coastline generally move in with their wave crests at an angle to the beach. As a result, the wave velocity is reduced at an uneven rate because one part of the wave reaches shallow water before other parts. The portion reaching shallow water first will also be the portion to slow down first. The change in the velocity of this part of the wave causes a change in its direction. This change in direction due to the different wave velocities is called *refraction* (Fig. 9-3). The refraction of the waves causes the waves to

9-3 Wave refraction: *(a)* concave; *(b)* convex.

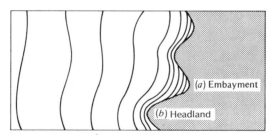

(a) Embayment

(b) Headland

turn toward the shallow water, so that the wave crests tend to parallel the depth contours.

Among the factors controlling wave refraction besides the water depth are the shape of the coast and bottom topography, the wavelength of the approaching wave, and the angle at which the wave approaches the shore. Along coasts where the shoreline is irregular, the wave energy is concentrated to produce convergence of the waves. This phenomenon occurs as the wave crests move over raised areas of bottom topography or as they approach a headland. In areas of depressed topography or where the waves enter a bay the wave energy is dispersed. The convergence of waves, on the other hand, concentrates the energy, which thus has a greater destructive effect on the coastline than diverging waves.

We have mentioned that a wave approaching the shore begins to feel the bottom when the water depth is one-half the wavelength. This causes the wave to break, forming what are commonly called *breakers,* when the depth is equal to 1.3 times the wave height. In general, there are three kinds of breaking waves, spilling, plunging, and surging (Fig. 9-4). The nature of the breaking wave is influenced by wave steepness, wind, direction, and bottom topography.

The spilling breaker tends to break gradually over an extended distance, covering a gently shoaling bottom. This form is usually the result of steep wind waves produced by strong winds blowing in the direction of wave travel. Plunging breakers form when wind is blowing against the direction of the wave advance and the bottom shoals rapidly. The plunging breaker begins to curl over and suddenly crashes all at once, often producing a booming sound. A surging breaker does not actually break in the same sense as the spilling or plunging forms; however, there is a noticeable peak that develops as this form moves up the face of the beach.

As the breaking waves rush onto a beach, some of the water continues up the slope until the force of gravity overcomes the landward motion, called the *swash.* The return of the water increases the tendency of the incoming waves to break.

The movement of the swash on the beach pushes or carries small particles of sand and shell material. When the backwash returns to the ocean, these particles are left behind, forming *swash marks,* indicating the highest point reached by a wave on the beach (Fig. 9-5).

Another type of wind-generated wave forms in storm regions or areas of strong wind. Waves which are produced in an area directly affected by the wind are termed *sea.* The sea is an irregular area hav-

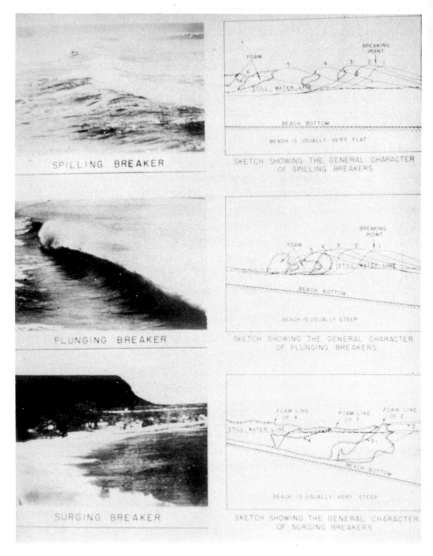

9-4   Three types of breakers: *(a)* spilling, *(b)* plunging, and *(c)* surging. Both photographs and diagrams of the three types of breakers are presented above. The sketches consist of a series of profiles of the waveform as it appears before breaking, during breaking, and after breaking. The numbers opposite the profile lines indicate the relative times of occurrences. *(Coastal Engineering Research Center, U.S. Army)*

9-5 A swash line along a typical beach. *(NOAA)*

ing a complicated mixture of long and short waves. Sea waves have a variety of periods and heights traveling in many different directions. As the waves leave the source area, the longer waves, which have a high velocity, outdistance the shorter waves with slower periods. As waves of similar size group together, the pattern of the waves becomes more uniform. The long, regular waves which continue to move after they have left the wind-generating area are termed *swell*.

OTHER TYPES OF WAVES

Internal Waves
Boundary, or internal, waves occur in stratified waters. The wind-generated waves we have just discussed occur at the boundary between the atmosphere and the water surface. However, differences in density with depth may produce waves which are not visible from the surface even though the internal waves may be as common as the wind-generated surface waves. These waves travel more slowly than surface waves but develop greater wave height. The highest recorded internal wave was 100 meters (300 feet) compared to 30 meters (100 feet) for a wind-generated wave. In addition to water masses of dif-

ferent densities, internal waves may be produced by mixing water of different types, which occurs when freshwater from the land rivers is fed into the oceans.

Stationary Waves

Besides progressive waves, there is another form of wave which does not exhibit forward movement of the waveform. These *standing waves* form only on the surface of confined bodies of water. The water surface remains stationary at the nodes while the rest of the water surface moves up and down (Fig. 9-6). The phenomenon is called *seiche* when it occurs on bays and lakes. A sudden storm or a rapid change in the atmospheric pressure produces disturbances on the water surface. Once set into motion, the water body continues to move, its oscillation being controlled by the length and depth of the basin (see Fig. 9-7).

Storm Waves

These catastrophic waves often accompany intense storms, the strong winds of which cause the water to pile up along the coast. When coupled with the usual high tide, the level of the sea becomes abnormally high, producing a dangerous situation for people in low-lying areas along the seacoast. These waves, which are produced by the combined forces of high winds and low atmospheric pressure, have been known to reach a height of 7 meters or more. Today, the prediction of hurricane in these low-lying districts results in the evacuation of the people to prevent a repetition of the disaster that struck Galveston, Texas, in 1900, when a storm surge rushed onshore drowning over 6,000 people. Storm surges produce a gradual increase in the level of the sea rather than the rhythmic rise and fall typical of other waveforms (Fig. 9-8).

9-6   A stationary waveform.

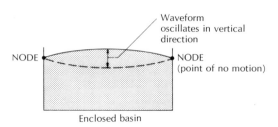

9-7 A ship tossed by high seas. *(U.S. Coast Guard)*

Tsunamis

The tsunami, or seismic sea wave, is the result of a large-scale submarine disturbance. The waves are set in motion initially by the sudden displacement of large sections of the Earth's crust beneath the ocean. Submarine earthquakes or submarine volcanic eruptions have enough energy to stimulate a series of large waves capable of moving thousands of miles across the sea. When the catastrophic disturbance occurs, a series of waves is sent out from the source area as a pattern of concentric circles, like those formed when a pebble is dropped into a still pond. The waves are exceptionally long, sometimes having a distance of as much as 160 kilometers (100 miles) or more between successive crests. While the wavelength of a tsunami is long, the wave height is very small, often as small as a few centimeters from trough to crest. Since this means that the waves have an extremely small steepness ratio, ships in the open ocean can pass right over a tsunami without even realizing it. However, when the wave reaches the shallow water, the water begins to pile up and may reach a height of 30 meters (100 feet). Waves of this magnitude have

9-8   Sea swell. *(NOAA)*

been known to move onto low-lying areas and affect people living as much as 2 miles from the coastline. A particularly devastating feature of seismic sea waves is the rate of speed at which they travel. Tsunamis may travel at speeds of over 400 miles per hour, and because in most cases the wave is not visible until it reaches shallow water, the moving wall of water may rush onshore without any warning. Seismologists have established a warning system to predict the arrival of these destructive waves and allow the inhabitants to move to higher land, thus reducing the loss of life if not the property damage (see Fig. 9-9).

COASTLINE FORMATION

The energy of waves moving onshore enables them to deposit sediments obtained through wave erosion to form the most common wave-deposited feature, the beaches. In general, all coastal features may be classified as being the result of either wave erosion or wave deposition. All features of the coastline are temporary and subject

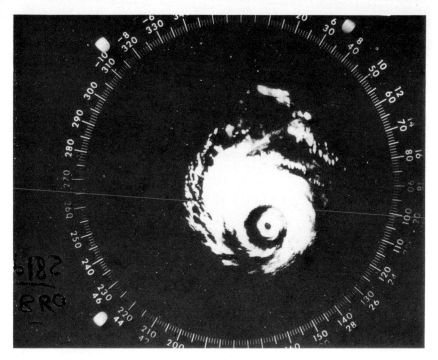

9-9 A radarscope of hurricane Beulah approaching Brownsville, Texas, in 1967. *(ESSA)*

to constant change in response to the energy expended by the sea when it meets the land. Several attempts have been made to classify shorelines but the most universally accepted scheme was proposed by Francis P. Shepard in 1937 and revised slightly in 1963. Essentially, Shepard's classification has two subdivisions, primary coasts and secondary coasts. The former are composed of terrestrial material while the latter are shaped by marine processes.

BEACHES

Beaches are formed by sediments which are transported by waves and tides and piled up along the shore (Fig. 9-10a). The sediments found on the beach are transported by the wave energy in a longshore direction as well as back and forth in rhythm to the movement of swash and backwash. Beaches tend to be in a constant state of flux, the observable features changing from season to season and in some places even daily.

The beach consists of the zone of wave-washed sediment extending along the coast from the low-tide level to the upper reaches of the shore affected by wave action. Although many of these terms are used interchangeably, most of the descriptive terms for the beach topography are quite specific (Fig. 9-10*b*). The *shoreline* represents the area where the land meets the sea while the term *shore* refers to

9-10   *(a)* A typical beach. *(Scripps Institution of Oceanography)*
(b) A cross section of an idealized beach. *(After Shepard)*

(a)

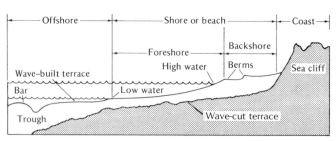

(b)

the zone between the mean low tide and the inner limit of wave-transported sand. The coast is generally considered the broad zone landward of the shore and includes all the topographic features related to waves and formed by their action.

Beaches are composed predominately of sediments the size known as *sand,* according to the geologist's scale of sediment sizes called the *Wentworth scale* (Table 5-1). Most continental beaches are composed of quartz sand both because it is abundant and because it offers extreme resistance to weathering, although many beaches consist of sediments of other minerals and even particles as large as pebbles and cobbles. The color and composition of beach sand vary from one part of the world to the next.

As the sediments pile up on the beach, the steepness of the beach is determined by the size of the detrital particles. In general, coarse sediments form steep slopes while the finer particles form wide flat beaches. The wide flat region of a beach is called the *berm.* It is quite possible to recognize more than one berm on the exposed section of a beach. This phenomenon results when from time to time waves which are higher than normal break against the shore to deposit sediments at levels higher than normal.

The *foreshore* is the sloping area over which the swash and the backwash transport their sediments. Sand removed from the shallow depths is transported and deposited on the foreshore, only to be removed again as the backwash of water returns to the sea. Generally, a *longshore bar* forms at the point where the breakers are developed. These bars are an accumulation of sediments that have been eroded from the berm and deposited in the breaker zone. A seasonal variation can be observed in some areas due to changes in the size of the waves. During the summer season, low waves transport the sediment shoreward, thus reducing the formation of the bar, while during the winter the waves are higher, beach sand is eroded seaward, and the bar forms offshore. When a bar is formed, it is normally separated from the surf zone by an intervening trough. During the summer season, the trough is filled in by the transported sediment as the bar is eroded by the action of the waves passing over it.

Waves and wave-induced currents that approach the shoreline at an angle produce a longshore transport of sediment. A movement of sediment occurs parallel to the beach, resulting in the transport of sediment along the beach. The longshore drift of sand may produce a *spit,* an elongated area of sand or gravel that projects from the mainland. It is the result of deposition of sediments that were transported by wave-created currents. As more and more sediment accumulates, the spit may lengthen enough to block the mouth of a bay completely. This barrier is the result of sediment accumulation in excess of the erosion due to currents moving into and out of the bay. When currents are able to remove sediments faster than they accumulate, the entrance to the bay remains open.

If the amount of available sediment is great, the beach area is widened by the amount of wave-carried sediments deposited on the shore. This condition, termed *progradation* as opposed to erosion, is uncommon. For the most part, the sediments of the beaches are eroded and transported to be deposited elsewhere. Although progradation appears to be a relatively rare feature of coastline formation today, there is ample evidence to suggest that it was particularly active in the geologic past.

Besides transporting sediments, the waves have enough energy to alter the consolidated features of the coastline. Waves breaking against the solid rock material of the coastline exert forces so strong that they may even alter the bedrock. The hydraulic pressures are able to push or pull large segments of rock from the consolidated mass. Furthermore, large rock fragments carried by the rushing water can shatter rock sections when they are flung against the rock wall by the surge of water.

In addition, the sediment-laden water wears away the rock materials with which it comes in contact. This abrasive action of the waves results in a smooth, rounded surface as well as a reduction in the overall size of the rock. Because of the rate of differential erosion due to the characteristics of the parent rock material, abraded wave-cut features are often quite dramatic. The result is a scalloped and irregular design in the rock material because weaker materials erode more quickly than more resistant forms. The formation of a wave-cut bench or cliff and a wave-built terrace is typical of the type of geologic structure that results from wave attack along a coastline. The cliff and the bench both result when erosion at the base of the cliff weakens and removes the underlying materials along the coastline. Eventually, the overlying materials collapse, widening the bench. The debris removed from the wave-cut structure is deposited seaward of the wave-cut bench to form a wave-built terrace. Through this process, as well as others, the coastline features are constantly subjected to change. Since the forms of coastline vary from place to place and their origins differ widely, a system of classification was needed before they could be subjected to an orderly investigation.

DEVELOPMENT AND CLASSIFICATION OF COASTLINES

In general there are two possible types of changes in the level of the sea. Either the water level itself will rise or fall, or the land may be uplifted or depressed. In addition, in respect to developmental processes of coastlines, these structures also have two methods of formation. Coasts are produced either by terrestrial means or by marine processes.

In his classification (modified in 1963) Shepard names two broad areas of division with many subdivisions listed beneath them. He bases his concept of primary and secondary coastlines on the mode of formation of the coasts. The primary coast is formed by land-based agencies, and the secondary coastlines are the result of marine factors (see Fig. 9-11).

9-11   The coastline classification scheme as developed by Shepard. *(After Shepard)*

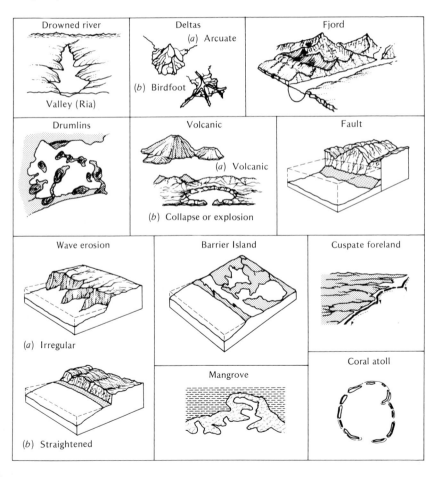

Primary Coastlines

Shepard's first subdivision of his primary coastline is the result of erosional processes that were subsequently drowned by the melting of glacial ice or the downwarping of the local landmass. As examples of this form he cites the ria coast (drowned river valleys), drowned glacial-erosion coast, such as fjord coast, and drowned karst topography (resulting from groundwater action). Shepard then groups together all coastlines formed as a result of depositional deposits. In this category he distinguishes the agents of deposition in the form of rivers, glaciers, and winds. The third category of land-established coastlines includes known volcanic eruptions of both the catastrophic type and the quiet flow of lava. The last form of terrestrially developed coastlines is produced by movements of the Earth's crust (diastrophism).

Secondary Coastlines

Shepard's classification of secondary coastlines consists of those formed chiefly by marine agencies or organisms. It is also possible that the secondary forms of coastlines were primary forms before being subjected to the modifying factors of the sea. The first subdivision in this category contains all those coastlines which have been formed as the result of erosion due to wave action. His second subdivision includes all varieties of coastlines formed as a result of marine deposition. In general, all these forms are due to beach progradation by waves and currents and produce coastlines that are easily recognized, some examples being the formation of barrier beaches, spits, tidal mud flats, and salt marshes. The third variety of marine-formed coastline develops when the marine organism itself is responsible for building the coast. The best-known example of this subdivision is the coral reef so common in the tropics.

Although widely accepted as a means of classifying coastlines, the Shepard classification has been subject to some criticism, at least in some quarters, due to his omission of specific forms, particularly his intentional omission of coastlines of emergence. In order to fully explain the concept of coastline classification some coastal forms omitted from the Shepard discussion will be covered at this time.

Coastlines of *emergence* occur when the Earth's crust rises near the border of a continent. The same effect can be observed if the sea level is reduced in relation to the land. As a result of these changes, wave-cut cliffs and wave-built terraces are raised to positions well above the area susceptible to further erosion by the action of the waves. In many areas, this process repeats itself, to form a series of elevated shorelines in a steplike arrangement. This is common in many places along the California coastline.

Other forms of coastlines not represented in the Shepard classification include neutral, fault, and compound coastlines. In a *neutral
coastline* new land is being built out into the water. The term neutral
is applied because there has been no noticeable change between the
sea level and the continental coastline. The *fault coastline* is unique
in being formed by massive crustal movement of the Earth that results
in a downdropping of a portion of the crust on the seaward side of
the fault. The landward block rises up, and the waves now break
against the fracture zone, forming the fault coastline. The *compound
coastline* is a combination of several types. For example, if the conditions necessary for forming a coastline of emergence are followed by
those necessary to produce a coastline of submergence, a compound
coastline will result.

SUMMARY QUESTIONS

1. Describe wave speed, wavelength, and wave height. How is each
   calculated?

2. Explain particle motion in waves.

3. How do breakers form, and what factors control breaker formation?

4. What are internal waves, and how are they created?

5. Describe coastline classification according to the Shepard
   system. What other system also is used?

6. What is differential erosion? What kinds of features are formed on coastlines as a result?

7. How do waves differ in their action on the shoreline from one season to the next?

8. What is the primary force producing waves across the open sea? What other factors produce waves?

9. How does particle motion in waves change with depth? To what depth is this motion effective?

10. What causes wave refraction and how does refraction affect wave motion?

11. What are swash and backwash? Are there any indications of either movement on the beach?

12. How does the beach form vary from winter to summer? What is the primary reason for the variation?

13. What coastline classifications are not covered by the Shepard system?

14. What are coastlines of emergence and submergence? How do they differ from the Shepard system?

15. Describe some of the depositional forms that one finds along various coastlines. What is the origin of the building materials for these features?

16. What effect will waves have on an organism deeper than half the wavelength of the wave?

17. What are swash marks? How are they created?

18. What are sea waves? How do they form?

19. What is a seiche?

20. What is the Wentworth scale? For what purpose is it used?

21. Define berm.

22. How do spits form?

23. What is progradation?

24. What is a fault coastline? How does it form?

25. What is the effect as a wave begins to "feel" the bottom?

26. What is a standing wave and how does it form?

27. What is a ria coast? Karst topography? Fjord coast?

10 | MINERAL RESOURCES OF THE SEA

As natural resources rapidly disappear from the land, man has turned to the oceans as his new source of supply. Food, minerals, even drugs, will one day be extracted from the waters of the ocean and the ocean floor. For as our population increases, the raw materials needed for our civilization are being used up more rapidly than ever before.

At the present rate, the world's population will be nearly double in the year 2000; about 7 billion people will inhabit the Earth. The population in the United States will increase from the current 200 million to nearly 330 million. And the countries of Asia, Africa, and South America will be industrialized just as those of North America and Europe are today. This means that natural resources will be needed in quantities more than double those used today.

Several studies have been made dealing with the world's future need for raw materials. In 1967, the United States alone used 1 ton of iron per person. Our need will be 12 times as much in the year 2000. Our need for copper will be about 12 times today's need (see Fig. 10-1a); and we shall require about 16 times as much lead.

The world is now in short supply of many resources which have been depleted from the landmasses. Mercury, used in temperature- and pressure-recording machines, is already in short supply. Tin, used in making alloys of many metals, is scarce. Photographic chemical manufacturers, the electronics industry, and battery and jewelry manufacturers require large amounts of silver, a metal very scarce today. Silver has become so scarce and

Manganese nodules. *(Photograph published with the permission of the Woods Hole Oceanographic Institution)*

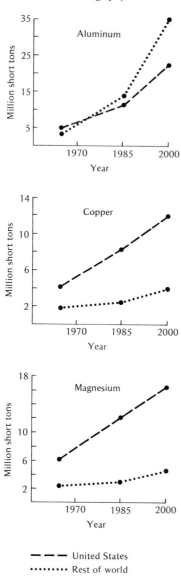

10-1   (a) The future requirements of some selected mineral resources.

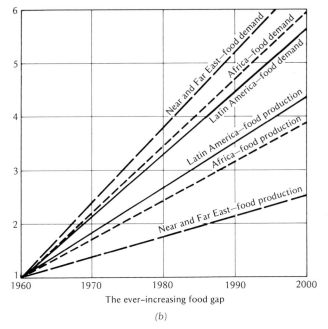

The ever–increasing food gap

(b)

10-1 (b) Present and future world food needs.

___

expensive that countries are removing it from all coins and medals. Cobalt, another metal in short supply, is needed for the manufacture of permanent magnets.

The oceans cover a surface area of 140 million square miles. There are millions upon millions of dollars worth of these natural resources in the oceans and on the ocean floors. Yet because of the pressure and depth we can examine only the continental shelves, which are immediately adjacent to the continents. This represents a small fraction of the total surface area of the oceans. For example, the United States defines the shelves as the area which is less than 656 feet deep or 50 miles out from the continents, whichever is greatest. Unless we develop international agreements and the means for exploring the ocean floor under international treaties, our sources of raw materials will always be limited and in short supply as we struggle to settle disputed claims for the sea floor.

In the next 35 years, the world will need more metal than has been used in the last 2,000 years of man's history. We shall need 3 times more energy in the next 20 years than we used in the last 100 years; and food production must be increased in proportion to our population growth. Where will all these raw materials and energy come from? They will have to be supplied by new sources, many of which will come from the sea. But the research will be demanding and expensive!

The United States government spent a total of 24 billion dollars for research in about 10 years before man set foot on the surface of the moon. Scientists believe that at least 20 billion dollars must be spent in the next 10 years on oceanographic exploration if man is to utilize the resources of the sea for his survival in the future.

THE RESOURCES OF THE SEA

How can we tell what natural resources are available on the ocean floor? Of course, no one really knows what the floor contains or in what quantities. However, there is good reason to expect that all we shall need can be obtained from the ocean.

Scientists believe that many of the elements found as part of the continental rocks will be found in the rock masses underlying the ocean basins. The ocean basins are a different area of the Earth, but in places they are composed of the same materials as the continents. Because the ocean covers such a vast area of the Earth, it is reasonable to expect that more of the same types of raw materials will be found in them (Fig. 10-2).

Ocean water has been analyzed and examined many times, resulting in a general formula for artificial seawater. In fact, some hobbyists raise salt water tropical fish using artificial sea salts in their aquariums, and the fish thrive.

One cubic mile of ocean water weighs over 4 billion tons. A cubic mile is an area 1 mile long, 1 mile wide, and 1 mile deep. If the materials dissolved in this water could be removed and treated, a great number of substances could be recovered. Approximately 150 million tons of salts would be contained in this water and 8 million tons of magnesium salts. And at least 25 tons of gold and 45 tons of silver would be found in the water. All this in only 1 cubic mile of ocean water! And there are 328 million cubic miles of ocean water covering our planet.

Many places in the world use common table salt recovered from ocean water. Every year 5 million tons of salt removed from ocean water is used.

As long ago as 1933, the Dow Chemical Company built the first plant for the extraction of chemicals from the sea. A plant built in North Carolina was designed to remove bromine from the seawater.

We also use iodine from the sea. Many substances, such as iodine,

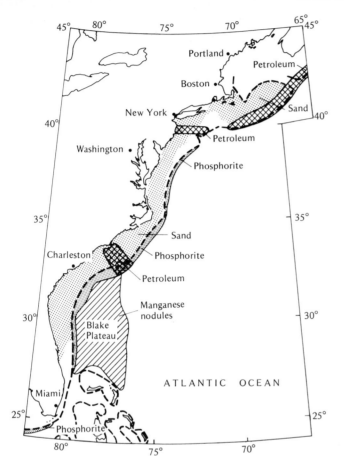

10-2 Some potential sites for mineral resources on the eastern continental margin of the United States. *(After U.S. Geol. Surv. Prof. Pap. 525-c)*

are found as minute traces in the seawater. Since these trace elements are impossible to extract directly from the water, science has to find some means of obtaining them indirectly.

Biologists discovered long ago that certain plants remove iodine from the water. The traces of iodine found in the water are concentrated in larger amounts in the cells and tissues of these plants. Seaweed, which concentrates iodine in its tissues, is our major source of iodine today. The seaweed is hauled out of the sea, and the plant is dried and burned. The iodine is extracted from the ashes. With proper treatment, the iodine can then be used in table salt and other substances and find its way easily into man's diet to prevent goiter. It may be possible someday to extract other trace elements from the sea through secondary sources such as seaweed or other organisms which concentrate the elements through their own particular metabolic activities.

One source of minerals from the ocean is manganese nodules. Manganese (with other elements) often settles out of solution and forms a deposit around some small object such as a shark's tooth on the sea floor. These nodules originate as small particles which increase in size to several inches as additional material settles out of solution. The sea floor and shelves are littered with millions of these nodules (Fig. 10-3). The nodules also contain small quantities of iron (14 percent), nickel (1.0 percent), copper (0.5 percent), and cobalt (0.5 percent). Someday, when the means of economically removing these nodules from the ocean floor has been developed, they may be the most valuable source of some of these elements in the sea.

10-3 A nodule deposit as shown on a television monitor aboard the R.V. *Prospector. (Photo by B. J. Nixon, Deepsea Ventures, Inc.)*

Manganese nodules are also found in other places. In 1968, millions of dollars worth of manganese nodules were discovered on the floor of Lake Michigan, one of the Great Lakes. Perhaps other enclosed basins will reveal large deposits of manganese nodules. Thus, even an inland waterway and its surrounding community may benefit from the knowledge we gain from oceanographic exploration.

At present, the most exciting discoveries of natural resources in the oceans have been oil-field explorations. It is possible to drill oil wells in the continental shelves, and the oil is obtained more easily than any other natural resource (Fig. 10-4). Many of the techniques used have been developed by scientific explorations of the ocean floor in the search for geologic samples of rock. These drilling techniques and their special devices have been combined with the knowledge developed by experienced oil-well operators (see Fig. 10-5).

The first offshore oil well was drilled in 1896, although large-scale offshore oil fields are a recent development. Oil companies have been searching for offshore oil pools on the continental shelves since the early 1930s. Scientists believe that at least 15 to 20 percent of the world's oil reserves lie under the sea. At present 6.5 million barrels of oil are recovered each day from offshore sites, or only about 6 percent of all the oil used in the free world. As the free supply of oil slowly dwindles, the search for offshore oil is spurred on by the great demand for new oil and even natural-gas supplies.

In 1968, oil and gas were discovered in the Gulf of Mexico at a depth of 11,753 feet. This was the first evidence of oil located in the ocean floor and not actually on the continental shelves. *Glomar Challenger,* an

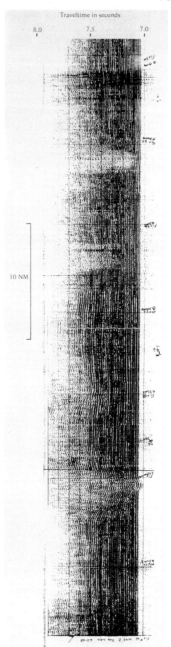

Traveltime in seconds

10 NM

10-4 A seismic profile showing a salt dome. *(Texas Instruments, Inc.)*

10-5   A deep-sea drilling vessel. *(Standard Oil Company of New Jersey)*

oceanographic research vessel, was studying a group of strange mounds known as the *Sigsbee knolls.* The ship is operated by the Scripps Institution of Oceanography and is part of the deep-sea drilling project discussed earlier.

One of the projects assigned to the *Glomar Challenger* was the recovery of drilled cores from the Sigsbee knolls. In a 450-foot drill core removed from the knolls, traces of oil and sulfur were noted. Although the object of the deep-sea drilling project was to obtain information about the history and origin of ocean basins, this discovery has commercial importance as well as scientific value. The presence of oil in the knolls indicates that they are actually salt domes which contain deposits of oil and gas. These concentrations of salt are often trapped beneath layers of rock in areas rich in oil deposits (Fig. 10-4).

At present, there are about 12,000 offshore oil wells in the world, and about 1,400 new wells are drilled each year. Almost all are directly on the continental shelves and are not actually drilled into the ocean floor.

In the same year as the discovery made by *Glomar Challenger,* another oil discovery was made off the north coast of Alaska. Several American oil companies cooperated in a joint exploration of the area and discovered one of the richest oil fields in the world. Estimates of the size of the field are that at least 5 billion and perhaps as much as 10 billion barrels of oil may be in this region.

MINING THE OCEAN FLOOR

In 1958, a conference dealing with the ocean floor and its uses was held in Geneva, Switzerland. At that time an international agreement was signed which gave the right to exploit the continental shelves to those countries which were bordered by the shelves. The United States, with this one agreement, acquired more land than all the land it has acquired since the Louisiana Purchase in 1803. The total area of the continental shelves bordering the United States is 800,000 square miles. Exploration and mining of the minerals on these shelves is still difficult but not impossible once adequate research has been carried out.

The United Nations has explored the possibilities of oceanographic mining and control of the ocean basins. There have been several proposals to turn control of the ocean floor and its resources over to the United Nations. It is felt that the minerals extracted could be the main support of the United Nations, instead of the contributions and dues now paid by each member country. Of course, most oceanographic and mining companies oppose this suggestion. They feel that each country should be allowed to explore and exploit the best it can.

Although recovery of the natural resources of the ocean floor may not be far away, no one can yet predict when that day will be. Almost all scientists agree that exploiting the ocean-floor resources requires new techniques. Some major scientific breakthrough in mining and drilling techniques will be needed before the oceans can be used as easily as the landmasses as a source of raw materials.

Although it is relatively easy to explore the continental shelves, it is not quite as easy to explore the ocean bottom. Nearly all oil wells presently operating are in depths no greater than 600 feet. Explorations and test drillings are now approaching 1,000 feet. These depths are slight indeed compared with the average depth in the ocean of 2½ miles.

Minerals must also be evaluated before oceanographic companies can determine whether it is practical to attempt to mine them. At the present time, the greatest practical depth at which minerals can be economically recovered is about 3,300 feet. In 10 years we may have the ability to utilize minerals from depths about 5 times those of today.

Several new techniques are already being used to extract the minerals from the ocean floor. Oil wells in the oceans are controlled by remote-control devices, and observations of the drilling can be made by specially constructed television cameras on the oil rigs.

Exploratory drilling and sampling techniques have allowed scientists to estimate the amount of manganese nodules on the ocean floor. It is believed that there are at least 200 billion tons of the nodules scattered about the ocean depths. One proposal for obtaining these nodules involves the use of a type of underwater vacuum cleaner. Tubes dropped from ships would scoop up the bottom sediments and nodules by a powerful suction device. The nodules would be gathered and the sediment returned to the sea (Fig. 10-6).

Underwater cameras and television have been used by archaeologists in search of half-buried objects. Many statues, ships, and other remnants of past civilizations have been recovered from many parts of the sea. Most of us think of undersea exploration as involving treasure hunters. Ships laden with gold, silver, and jewelry have been discovered in this way, of course. But explorers have also discovered jars of oil, wines, food, and many other staples used to outfit the long voyages made by the ancient mariners (Fig. 10-7).

Mining of a few minerals already is being carried on in different parts of the world. The continental shelf off the coast of southwestern Africa has yielded diamonds, washed by rivers from the volcanic rock containing them in the interior of the continent.

Recently, large deposits of tin were discovered around Thailand, Malaysia, and Indonesia. When a practical means of mining this mineral has been developed, it will be very important to countries such as these which are just becoming industrialized. A few deposits of minerals are currently being recovered by dredging operations

10-6 R.V. *Prospector*, length 152 feet, beam 32 feet, 6 inches. *(Deepsea Ventures, Inc.)*

in some parts of the world since this technique is inexpensive in shallow areas.

Explorations off the coast of Alaska have proved that platinum is present in the continental shelves. In addition, there is evidence that important deposits of gold, copper, and diamonds may also be found there.

Englishmen heat their homes and industrial furnaces with coal recovered from the sea. There are several tunnels running under the margins of the North Sea, which are rich in coal deposits. The tunnels are started on land and run under the sea through the coal deposits.

In addition to these minerals, the ocean has important deposits of materials for construction purposes. Sand is obtained from coastal areas for the construction of large buildings. And the shells of marine organisms contain carbonate, another important building material used for the lime in cement. Each year, the United States uses some 20 million tons of shells recovered from the continental shelves, in addition to several million tons of gravel. One recently discovered

10-7   Research submarine *Asherah* and diver. *(General Dynamics, Inc.)*

source off the coast of New Jersey is estimated to contain 1 billion tons of valuable gravel deposits. These materials can be obtained with relative ease by dredging operations (Fig. 10-8).

THE TREASURE OF SEAWATER

Although seawater contains a great many elements, their recovery is extremely difficult because they are present in such small amounts. It may be possible one day to extract iron, silver, and mercury in the same way we now obtain iodine from seawater. Some indirect means may allow us to obtain these metals cheaply. At present, with one major exception (magnesium), metal recovered from seawater remains a dream for the future. We now extract all the world's supply of magnesium from ocean water. In the United States alone this process represents a 70 million dollar annual enterprise.

Other minerals are not easily extracted from ocean water because their quantity in solution is low. For example, seawater contains

10-8 A dredge.

enough gold to give 25,000 dollars worth to each person in the United States. However, the cost of obtaining the gold from the water would be almost 50,000 dollars—hardly a good bargain!

However, some advances are being made in the extraction of rare metals from seawater. In England, a group of scientists has devised such a means of extracting uranium. The cost of this uranium is about 20 dollars per pound. At the present time, uranium costs around 12 to 15 dollars per pound. As the cost of extracting the uranium drops and uranium on the continents becomes scarcer, this may become a worthwhile venture.

The most valuable resource in the water of the ocean is the water itself. As the world's population increases, new supplies of water are needed. And the sea will be the source of supply for future generations. Already many techniques are in operation in various parts of the world to obtain drinking water from the ocean (Fig. 10-9).

In 10 years, a larger portion of the world's supply of freshwater may come from the sea. The water will be either desalinated (salt removed) or come from freshwater artesian wells in the sea floor. Recently, such natural freshwater wells were found off the coast of Florida.

One suggestion for new supplies of freshwater has been to use icebergs. It is possible that one day icebergs may be towed from polar

(a)

(b)

10-9 Desalting projects: *(a)* the Soviet Union; *(b)* Israel. *(FAO-UN)*

regions to population centers near the coastlines. The icebergs, which are almost pure water, would be melted and the water used for drinking purposes.

Much experimentation is presently being carried on into the development of desalination plants, which use a variety of processes to remove the salts from ocean water. The purified, desalted water can then be used for human consumption.

A plant in the Florida keys has been producing desalted water for several years. The cost of the water is about 85 cents per 1,000 gallons. When the plant first went into operation, the cost was close to 4 dollars per 1,000 gallons. It is hoped that the cost will reach about 35 cents, which is about the cost of water from land sources. This plant is just one of nearly 700 such plants currently desalting 25,000 gallons of drinking water daily throughout the world.

The methods of desalting vary from place to place. Among the techniques being investigated are the use of solar energy to evaporate the water. In another technique, flash distillation, the water is pumped into heating chambers under reduced pressure. Thus, the water is vaporized at temperatures much lower than the usual 100 °C (212 °F) under 1 atmosphere of pressure. This technique reduces the amount of energy necessary for the desalting process.

One of the greatest problems in producing cheap drinking water from the sea is the cost of the heat energy. Most of the ways in which water is desalted use heat to change the water into a gas, which leaves the salt behind. Some scientists believe that the energy wasted in electric power plants might be used instead to heat seawater directly. Thus, a plant might produce electricity and drinking water at the same time for a nearby city. Japanese scientists are presently investigating the possibility of building such plants.

The other method of saving money on the cost of producing desalted water is to sell the by-products from the salt removal. If a desalting plant could produce several billion gallons of drinking water a year, it could sell the by-products to a variety of manufacturers. From this water, the plant could obtain about 100 million tons of salt, 8 million tons of magnesium, 2 million tons of potash for fertilizer, 250,000 tons of bromine, and 6 pounds of gold. The sale of these minerals would help lower the cost of producing the drinking water. Many other minerals could also be obtained to further lower the cost of the water.

We are in a period of intense oceanographic investigation known as the International Decade of Oceanographic Exploration. During this operation, all the major maritime nations of the world are cooperating in a massive examination of the world's major water bodies in an effort to gain a greater understanding of the oceans and their potential as sources of supply.

Recently, a series of hot brines was discovered in the middle of the Red Sea floor. The waters, heated by magmatic activity below the sea floor, have become brines in which minerals have concentrated up to 300,000 parts per million, about 10 times the average salinity of such water. In these brines are concentrations of heavy metals. In addition, the sediments beneath the brine are rich in heavy metals: gold, zinc, copper, silver, and lead. Basins like the enclosed Red Sea may prove to be a valuable source of economically worthwhile concentrated brines.

Thus, although ocean water is rich in minerals, man is still far from the day when he will be able to use the minerals in the water. But the experiments go on, and our knowledge slowly increases about the sea and its potential. The day may come when we eat algae bread, drink desalted ocean water, eat beef raised on food from the sea and spinach fertilized by minerals from seawater, and drive a car built entirely of metal recovered from the sea.

SUMMARY QUESTIONS

1. What are the two major sources of new raw materials from the sea?

2. How do we now recover elements such as iodine from the sea by indirect processes?

3. What important resources may be recovered from sources such as manganese nodules? Why are we currently prevented from exploiting this potential?

4. Why are the continental shelves particularly valuable oceanic property?

5. What materials from the sea are important for construction purposes? How do we recover these materials now?

6. How might desalting plants be made more economical by producing secondary products? Describe the present production of at least two different types of secondary products in modern desalting plants.

7. One problem associated with power plants is thermal pollution, the excess heating of river or lake water. How might this be eliminated by means of a desalting plant?

8. What definition is used to delimit the continental shelves? What is the amount of shelf now owned by the United States?

9. What metal is now totally recovered from sea sources? List at least one other important chemical now obtained almost exclusively by recovery from the ocean.

10. How much oil is calculated to lie under the sea floor? From what sea source and location are we currently recovering oil?

11. What do we believe the strange hills known as *knolls* result from? How were these facts uncovered?

12. Where have some continental shelves already yielded mineral resources? What resources have revealed themselves?

13. Briefly describe flash distillation and how this reduces the cost of desalting water.

14. What are hot brines and how are they significant?

15. Why are such ordinary substances as shells and sand considered important economic resources? What is their major use?

11 | MAN AND THE SEA

Although oceanographers have found it virtually impossible to agree on all details of what constitutes the ideal research vessel, they are unanimous in requiring a stable platform on which to work. The research vessel becomes the scientist's laboratory on the open sea, and stability is vital to the marine scientist because without it his instruments cannot function accurately. The instruments lowered over the side during experimentation must function as intended and remain at a particular location so that they are recoverable at the proper moment. In addition, instruments used on the vessel itself must be specially engineered to permit operation during all kinds of weather.

To develop a stable platform at the site of exploration, oceanographers have experimented with many different forms of vessel. Instead of concentrating solely on the development of surface craft, they have also investigated and used successfully many diverse types of submarine vessels. The development of bathyscaphes and deep-diving submarines has done much to advance the study of the ocean basins.

In an attempt to observe the ocean bottom topography directly man has ventured into an alien world. Man's physical and biological makeup does not equip him for life in the sea. As the compulsion to seek out and explore the unknown overcomes man, he has created devices that allow him to spend increasingly long periods of time beneath the ocean surface. As technological sophistication increases, permanent deep-sea stations will be established, where scientists can live and work for extended periods of time.

Trieste II. (U.S. Navy)

DIVING BENEATH THE OCEAN SURFACE

As mentioned in Chap. 1, legends link Alexander the Great with deep-sea exploration in a diving bell. Modern developments in underwater breathing apparatus began around 1830, when a diving helmet was attached to a closed suit and connected to an air pump, which allowed a man to explore the depths beneath the surface. Rudimentary as this equipment was, it was used successfully in depths as great as 200 feet. Surprisingly, the individual diving equipment of today has not changed substantially from that of 140 years ago (see Fig. 11-1).

By the middle 1800s, a self-contained diving gear had been devel-

---

11-1 A hard-hat diving suit. *(CERC, U.S. Army)*

oped. The origins of this gear are traced to a merchant seaman who was reported to have strapped two tanks of pure oxygen to his back with breathing tubes extending from the tanks. Although the first devices utilized pure oxygen, this technique was found to be dangerous beyond a depth of 30 feet, where it may produce convulsions resulting in unconsciousness and death. The use of compressed air (a mixture of oxygen and other gases) has eliminated part of this problem.

In 1943, Jacques-Yves Cousteau and a French engineer, Emile Gagnan, developed an aqualung with a regulator device that permitted a flow of air on demand regardless of the diver's position beneath the water. This device is called a *scuba,* an acronym for Self-Contained Underwater Breathing Apparatus. With the addition of rubber foot fins and goggles, man's limitations beneath the sea were reduced considerably (Fig. 11-2).

The aqualung allows air to be delivered to the diver at the same pressure that is being exerted on his body by the water. The tissues of the human body are denser than the water so that as long as the air being delivered to the diver's lungs is kept at the same pressure as the body, the weight of the water has little effect on the body. The air in a scuba tank is compressed to 200 atmospheres and supplied to the diver on demand. As the diver breathes in on the mouthpiece held between his teeth, a stream of air is supplied to his lungs, and as he exhales, the air is expelled through the regulator. The inhaled air is separated from the exhaled air by a rubber diaphragm in the regulator housing.

The limitations of this equipment center around the depth to which one can descend and the amount of time permissible at that particular depth (Fig. 11-3). The amount of gas normally found in human tissues at atmospheric pressure increases with an increase in depth. At a level in excess of 130 feet (although the phenomenon may occur at lesser depths), the amount of nitrogen in the tissues produces a nitrogen narcosis known as *rapture of the deep.* The increased level of nitrogen in the blood produces a euphoria similar to that produced by too much alcohol; that is, the diver loses his sense of reality and may do abnormal things such as taking off his face mask and offering air to the fish. In addition, the diver may lose his sense of up or down due to the increased pressure on the inner ear controlling his sense of equilibrium. Some divers have been known to plunge to their deaths by going deeper instead of toward the surface.

The rate at which a diver ascends or descends is extremely critical in preventing another dangerous condition commonly known as *the bends.* It may cause severe pains in the joints, respiratory difficul-

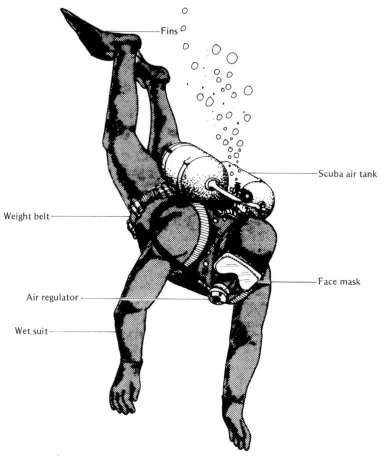

Fins

Scuba air tank

Weight belt

Face mask

Air regulator

Wet suit

11-2   Scuba gear. *(CERC, U.S. Army)*

ty, or nervousness, and in extreme cases it causes death. The bends occur when a diver surfaces too rapidly. The reduction in pressure frees gases from the tissues and forms bubbles, which lodge in the joints, blood vessels, and brain.

Since the amount of air available to a free diver is limited and the diver is often required to spend a specific amount of time doing work in the deep ocean, the U.S. Navy has developed a standard decompression table (Table 11-1) which dictates the amount of time the diver may safely stay on the sea floor and the amount of time necessary for ascent. For example, a 50-foot dive for a period of 78 minutes would be allowable without decompression, but for longer periods of time the diver would have to stop periodically to acclimate

(a)

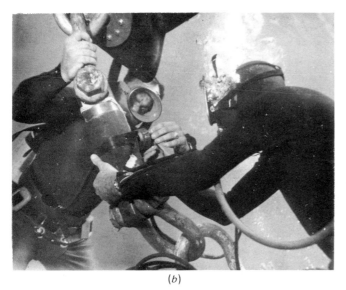

(b)

11-3 *(a)* Divers at work from the Tektite habitat. *(NOAA)*
(b) Divers working on underwater salvage. *(U.S. Naval Oceanographic Office)*

---

the body and prevent the bends. A dive to 100 feet without decompression is permissible only if the diver does not stay over 25 minutes at this depth. If the diver remained at the 100-foot level for 1 hour, he would require three stops between 30 feet and the surface to decompress. One problem of decompression is that the amount of air in the scuba tank is limited and with the necessary precautionary

TABLE 11-1   U.S. Navy Standard Air Decompression Table

Instructions for Use

Time of decompression stops in the table is in minutes.

Enter the table at the exact or the next greater depth than the maximum depth attained during the dive. Select the listed bottom time that is exactly equal to or is next greater than the bottom time of the dive. Maintain the diver's chest as close as possible to each decompression depth for the number of minutes listed. The rate of ascent *between* stops is not critical for stops of 50 feet or less. Commence timing each stop on arrival at the decompression depth and resume ascent when the specified time has lapsed.

For example—a dive to 82 feet for 36 minutes. To determine the proper decompression procedure: The next greater depth listed in this table is 90 feet. The next greater bottom time listed opposite 90 feet is 40. Stop 7 minutes at 10 feet in accordance with the 90/40 schedule.

For example—a dive to 110 feet for 30 minutes. It is known that the depth did not exceed 110 feet. To determine the proper decompression schedule: The exact depth of 110 feet is listed. The exact bottom time of 30 minutes is listed opposite 110 feet. Decompress according to the 110/30 schedule unless the dive was particularly cold or arduous. In that case, go to the schedule for the next deeper and longer dive, i.e., 120/40.

Depth (feet)	Bottom time (min)	Time to first stop (min: sec)	Decompression stops (feet) 50	40	30	20	10	Total ascent (min: sec)
40	200						0	0:40
	210	0:30					2	2:40
	230	0:30					7	7:40
	250	0:30					11	11:40
	270	0:30					15	15:40
	300	0:30					19	19:40
50	100						0	0:50
	110	0:40					3	3:50
	120	0:40					5	5:50
	140	0:40					10	10:50
	160	0:40					21	21:50
	180	0:40					29	29:50
	200	0:40					35	35:50
	220	0:40					40	40:50
	240	0:40					47	47:50
60	60						0	1:00
	70	0:50					2	3:00
	80	0:50					7	8:00
	100	0:50					14	15:00
	120	0:50					26	27:00
	140	0:50					39	40:00
	160	0:50					48	49:00
	180	0:50					56	57:00
	200	0:40				1	69	71:00

Depth (feet)	Bottom time (min)	Time to first stop (min: sec)	Decompression stops (feet)					Total ascent (min: sec)
			50	40	30	20	10	
70	50						0	1:10
	60	1:00					8	9:10
	70	1:00					14	15:10
	80	1:00					18	19:10
	90	1:00					23	24:10
	100	1:00					33	34:10
	110	0:50				2	41	44:10
	120	0:50				4	47	52:10
	130	0:50				6	52	59:10
	140	0:50				8	56	65:10
	150	0:50				9	61	71:10
	160	0:50				13	72	86:10
	170	0:50				19	79	99:10
80	40						0	1:20
	50	1:10					10	11:20
	60	1:10					17	18:20
	70	1:10					23	24:20
	80	1:00				2	31	34:20
	90	1:00				7	39	47:20
	100	1:00				11	46	58:20
	110	1:00				13	53	67:20
	120	1:00				17	56	74:20
	130	1:00				19	63	83:20
	140	1:00				26	69	96.20
	150	1:00				32	77	110:20
90	30						0	1:30
	40	1:20					7	8:30
	50	1:20					18	19:30
	60	1:20					25	26:30
	70	1:10				7	30	38:30
	80	1:10				13	40	54:30
	90	1:10				18	48	67:30
	100	1:10				21	54	76:30
	110	1:10				24	61	86:30
	120	1:10				32	68	101:30
	130	1:00			5	36	74	116:30
100	25						0	1:40
	30	1:30					3	4:40
	40	1:30					15	16:40
	50	1:20				2	24	27:40
	60	1:20				9	28	38:40
	70	1:20				17	39	57:40
	80	1:20				23	48	72:40
	90	1:10			3	23	57	84:40

TABLE 11-1 (cont.)

Depth (feet)	Bottom time (min)	Time to first stop (min: sec)	Decompression stops (feet)					Total ascent (min: sec)
			50	40	30	20	10	
100	100	1:10			7	23	66	97:40
	110	1:10			10	34	72	117:40
	120	1:10			12	41	78	132:40
110	20						0	1:50
	25	1:40					3	4:50
	30	1:40					7	8:50
	40	1:30				2	21	24:50
	50	1:30				8	26	35:50
	60	1:30				18	36	55:50
	70	1:20			1	23	48	73:50
	80	1:20			7	23	57	88:50
	90	1:20			12	30	64	107:50
	100	1:20			14	37	72	125:50
120	15						0	2:00
	20	1:50					2	4:00
	25	1:50					6	8:00
	30	1:50					14	16:00
	40	1:40				4	25	32:00
	50	1:40				13	31	48:00
	60	1:30			2	22	45	71:00
	70	1:30			9	23	55	89:00
	80	1:30			15	27	63	107:00
	90	1:30			19	37	74	132:00
	100	1:30			23	45	80	150:00
130	10						0	2:10
	15	2:00					1	3:10
	20	2:00					4	6:10
	25	2:00					10	12:10
	30	1:50				3	18	23:10
	40	1:50				10	25	37:10
	50	1:40			3	21	37	63:10
	60	1:40			9	23	52	86:10
	70	1:40			16	24	61	103:10
	80	1:30		3	19	35	72	131:10
	90	1:30		8	19	45	80	154:40
140	10						0	2:20
	15	2:10					2	4:20
	20	2:10					6	8:20
	25	2:00				2	14	18:20
	30	2:00				5	21	28:20
	40	1:50			2	16	26	46:20
	50	1:50			6	24	41	76:20
	60	1:50			16	23	56	97:20
	70	1:40		4	19	32	68	125:20
	80	1:40		10	23	41	79	155:20

Depth (feet)	Bottom time (min)	Time to first stop (min: sec)	Decompression stops (feet)					Total ascent (min: sec)
			50	40	30	20	10	
150	5						0	2:30
	10	2:20					1	3:30
	15	2:20					3	5:30
	20	2:10				2	7	11:30
	25	2:10				4	17	23:30
	30	2:10				8	24	34:30
	40	2:00			5	19	33	59:30
	50	2:00			12	23	51	88:30
	60	1:50		3	19	26	62	112:30
	70	1:50		11	19	39	75	146:30
	80	1:40	1	17	19	50	84	173:30
160	5						0	2:40
	10	2:30					1	3:40
	15	2:20				1	4	7:40
	20	2:20				3	11	16:40
	25	2:20				7	20	29:40
	30	2:10			2	11	25	40:40
	40	2:10			7	23	39	71:40
	50	2:00		2	16	23	55	98:40
	60	2:00		9	17	33	69	132:40
	70	1:50	1	17	22	44	80	166:40
170	5						0	2:50
	10	2:40					2	4:50
	15	2:30				2	5	9:50
	20	2:30				4	15	21:50
	25	2:20			2	7	23	34:50
	30	2:20			4	13	26	45:50
	40	2:10		1	10	23	45	81:50
	50	2:10		5	18	23	61	109:50
	60	2:00	2	15	22	37	74	152:50
	70	2:00	8	17	19	51	86	183:50
180	5						0	3:00
	10	2:50					3	6:00
	15	2:40				3	6	12:00
	20	2:30			1	5	17	26:00
	25	2:30			3	10	24	40:00
	30	2:30			6	17	27	53:00
	40	2:20		3	14	23	50	93:00
	50	2:10	2	9	19	30	65	128:00
	60	2:10	5	16	19	44	81	168:00
190	5						0	3:10
	10	2:50				1	3	7:10
	15	2:50				4	7	14:10
	20	2:40			2	6	20	31:10
	25	2:40			5	11	25	44:10

TABLE 11-1   (cont.)

Depth (feet)	Bottom time (min)	Time to first stop (min: sec)	Decompression stops (feet)					Total ascent (min: sec)
			50	40	30	20	10	
190	30	2:30		1	8	19	32	63:10
	40	2:30		8	14	23	55	103:10
	50	2:20	4	13	22	33	72	147:10
	60	2:20	10	17	19	50	84	183:10

time the diver can easily exhaust his supply of air. Thus a system of sending down fresh tanks is often employed and has worked successfully as long as the diver heeds the safety time of decompression.

A general rule followed in even the shallowest of dives is that a diver should never ascend faster than the air bubbles he emits. Since air expands as it rises, a rapid ascent to the surface could cause the air pressure building up in the lungs to rupture the lung tissues. Presently, experiments are being conducted into the feasibility of using other gases mixed with oxygen which do not cause intoxication when dissolved in the brain. Although this may eliminate the rapture of the deep, it would not eliminate the need to decompress. In addition, experiments are being conducted using liquids rather than gaseous compressed air to enable divers to spend longer and longer periods of time in deep water.

SUBMERSIBLES

Although many advances have been made in the science of scuba diving (such as rubber suits, which retain the body heat to keep the diver warm in deep water; underwater sleds, which propel the divers through the water; and underwater cameras, which allow the scientists to document their observations), descents to greater depths require deep-diving submersibles. These vessels, which may even be stationed on the sea floor, maintain an artificial environment designed to keep the inhabitants dry, warm, and relatively comfortable while maintaining a controlled atmospheric pressure. Some of these craft will be designed so that divers may come and go from them; others will serve as base stations with man living beneath the sea for extended periods of time. The diving capabilities of the world's research submersibles are varied, being designed in some cases for specific projects (see Fig. 11-5).

(a)

(b)

11-4   Beebe and the bathysphere. *(U.S. Information Agency)*

RESEARCH SUBMERSIBLES

One of the earliest developments enabling man to venture into the sea was the invention in 1930, by Otis Barton, of a steel diving sphere (Fig. 11-4a). This bathysphere, or deep sphere, was suspended from a surface ship by a steel cable. This vessel had its own air supply, but power and communication with the surface ship were by means of an electric cable. In 1934, William Beebe and Otis Barton descended to a depth of 3,028 feet, a record that held until 1948, when Barton descended in an improved model of the bathysphere to 4,050 feet (see Fig. 11-4b).

On February 15, 1954, Georges Houot and Pierre Willm of the French navy became the first men to reach the bottom of the deep ocean when they dived to a record depth of 13,284 feet off Dakar, West Africa. This dive was accomplished in a bathyscaphe, or deep-boat, developed by Auguste Piccard, the Swiss physicist and engineer (Fig. 11-6). Piccard, who was also a record-breaking balloonist, modeled his craft using the principle of his gas balloon. The cabin depth was maintained by suspension from a large flotation device on the surface. In order to descend the depths of the sea, he added iron

---

11-5 R.V. *Alvin*—a modern research submersible. *(U.S. Naval Oceanographic Office)*

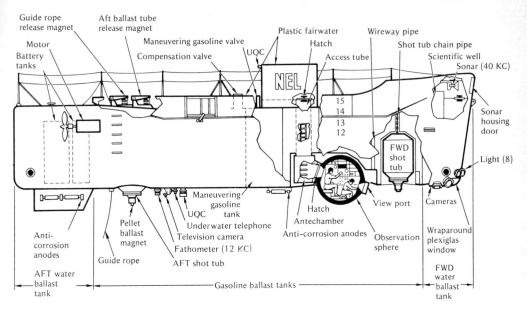

Guide rope release magnet
Aft ballast tube release magnet
Motor
Maneuvering gasoline valve
Compensation valve
Battery tanks
UQC
Plastic fairwater
Hatch
Wireway pipe
Access tube
Shot tub chain pipe
Scientific well
Sonar (40 KC)
NEL
15
14
13
12
FWD shot tub
Sonar housing door
Light (8)
Anti-corrosion anodes
Pellet ballast magnet
Guide rope
Maneuvering gasoline tank
UQC
Underwater telephone
Television camera
Fathometer (12 KC)
AFT shot tub
Hatch
Antechamber
Anti-corrosion anodes
View port
Cameras
Observation sphere
Wraparound plexiglas window
AFT water ballast tank
Gasoline ballast tanks
FWD water ballast tank

11-6   Cutaway sketch of *Trieste II. (USNOO)*

ballast which could be jettisoned for the return to the surface. The vessel was carried in the hold of a large vessel to the point in the ocean to be explored. The scientists sealed themselves in the pressure-resisting cabin, and the entire bathyscaphe was lowered into the sea.

In the spring of 1952, Piccard was commissioned by a citizens' group in the Italian city of Trieste to build a bathyscaphe in their city. Recognizing the problem encountered with building the first bathyscaphe Piccard accepted the commission, anxious to develop an improved model. The new model, which he named *Trieste II,* was filled with gasoline for buoyancy while in port and towed to the diving site, where the scientists could enter the cabin with the vessel in the water. This in itself was an improvement in the preparation time saved and in the operating convenience. In addition, the *Trieste II* was a stronger vessel than the prototype, but in all other aspects the two submersibles were very similar. The *Trieste II* was purchased by the U.S. Navy Electronics Laboratory in 1958. The plan was for the *Trieste* to dive to the deepest part of the oceans, the trenches. On January 23, 1960, Jacques Piccard, Auguste's son, and Don Walsh made this a reality by diving to the bottom of the Mariana Trench, some 35,800 feet deep (Fig. 11-7).

The Barton bathysphere had the inherent danger that if the cable

(a)

(b)

11-7   (a) *Trieste II* and its crew. *(U.S. Navy)*
(b) Jacques Piccard and Lt. Don Walsh, who first descended to a depth of 35,800 feet in the Mariana Trench. *(U.S. Navy)*

(a)

(b)

11-8   *(a) Deep Star. (USNOO)*
(b) Deep-ocean work boat (DOWB). *(General Motors Corporation)*

broke, the consequences would be fatal, but, the Piccard model would return to the surface as a float if a malfunction occurred. The float was designed in such a way that as it sank, seawater would compress the gasoline. This ensured that the pressure inside the float would be the same as the outside. This also saved money by allowing thinner metals to be used.

The iron ballast was held in place by a series of electromagnets. Should the vessel run into difficulty, such as a power loss, the electromagnets would automatically release, jettisoning the ballast and causing the float to bob to the surface like a cork. The ballast remained in the silos as long as current flowed through the electromagnets. When the current was turned off, the ballast flowed out freely. As the bathyscaphe sank, it picked up speed, so that ballast had to be dropped during descent to compensate for the loss of buoyancy.

The Piccards also developed a mesoscaphe, or middle-depth diving ship. Where their bathyscaphes were modeled after gas balloons, the mesoscaphe is an underwater helicopter.

Designed for use in waters up to 6,000 feet deep, the mesoscaphe uses light materials, thus reducing cost of construction. Since the mesoscaphe is lighter than water, if the propeller drive fails, the device floats to the surface by itself (Fig. 11-8).

The bathyscaphes represent the forerunner of the type of vehicles man will use to explore the bottom of the sea. However, some of the inherent limitations in the design of the bathyscaphes paved the way for more versatile devices.

An extremely maneuverable submersible, at present limited to the shallow area of the continental shelf, is the two-man diving saucer, developed by Cousteau, which can be used to explore submarine canyons and caves. Although the two-man model has been limited to depths of less than 1,000 feet, larger vehicles are being designed capable of diving to considerably greater depths.

One of the most impressive submarine research vehicles ever built is the deep-diving abyssal craft *Aluminaut* (Fig. 11-9). Named for the strong metal used in its construction and the Greek word for sailors, the *Aluminaut* was built by the General Dynamics Corporation for the Reynolds Metals Company. It is designed to reach depths of 15,000 feet. Built to resemble conventional submarines, it is self-buoyant, using water as the main ballast system, with an additional ironshot system similar to those in the bathyscaphe.

Although submersibles represent the exciting phase of deep-sea exploration, one must not neglect the importance of the special vessels and devices used in oceanographic research.

11-9 *Aluminaut,* the Reynolds Metals Company working deep-sea submersible, has a design capability of 15,000 feet. *(Reynolds Metals Company)*

A unique research vessel *Flip* (Floating Instrument Platform) is being operated by Scripps Institution of Oceanography at La Jolla, California (Fig. 11-10). *Flip* is a 355-foot-long vessel which can be towed to the research site in a horizontal position; there her ballast tanks are flooded, causing her to flip into a vertical position. This affords the scientists a stable platform from which to conduct their scientific studies. When the investigations are completed in one place, the water is blown out of the ballast tanks and the vessel returns to a horizontal position to be towed back to shore or to another site.

(a)

(b)

11-10 *Flip,* a 355-foot floating instrument platform: *(a)* horizontal position; *(b)* vertical working position showing 55 feet out of the water. Note the observation platform. *(Scripps Institution of Oceanography)*

(a)

(b)

11-11   Offshore drilling rigs. [*Standard Oil Company (N.J.)*]

Other types of platforms being used are anchored buoy systems, which are placed at specific stations to record weather conditions, surface or deep-water currents, and wave action. The system is anchored to the sea bottom with a flotation device at the surface. Along the line are several sensors, which record and transmit information to the instruments in the flotation buoy. In the more advanced equipment the data are transmitted directly to land-based recording stations (Fig. 11-11).

The use of platforms permanently fixed to the sea floor has also been adopted for oceanographic investigations. These platforms serve as radar and navigational towers; some have living quarters to enable the scientists to live in their deep-sea laboratory while experimentation is being conducted.

The U.S. Navy Sealab program consists of an underwater laboratory designed to test man's ability to live underwater for extended periods of time. In 1964, during the Sealab I project, four divers stayed at a depth of 193 feet for 11 days. In the follow-up study in 1965, the Sealab II project, three 10-man teams each spent a period of 2 weeks at a depth of 205 feet. The astronaut turned aquanaut, Scott Carpenter, who was in charge of the program, spent 30 days at that level.

Plans are in the works by the Navy for a Sealab III program (Fig. 11-12). Despite some setbacks, the plans call for five 8-man teams to live a total of 12 days each at a depth of 610 feet.

The U.S. Department of the Interior project Tektite is a cooperative multiagency program combining the efforts of industry and university scientists to develop a submarine habitat in which a team of scientists live and work for extended periods of time (Fig. 11-13). Tektite I was conducted near St. John, in the Virgin Islands, from February 15 to April 15, 1969. The system utilized a nitrogen-oxygen mixture in the controlled atmosphere. Constant monitoring of the physiological reactions of the scientists was carried on by a team of specialists at the surface. The program included daily experiments requiring the scientists to leave the habitat for varying periods of time each day.

As a result of the information gained through the Tektite I project a follow-up study was conducted in 1971. Tektite II includes 17 missions spread over a 7-month period. The missions are scheduled to last from 14 to 20 days each at varying depths of between 50 and 100 feet. In some cases there will be overlapping crews which will mean extended periods of saturation for certain members of the team. The hope is to train a large number of scientists in saturation diving methods permitting direct and continuous access to the marine environment.

(a)

(b)

11-12 (a) Sealab III habitat.
(b) An artist's conception of Sealab III habitat with support vessel. *(U.S. Navy)*

11-13   Tektite I habitat: cutaway sketch of the facilities of Tektite. *(U.S. Navy)*

OCEANOGRAPHY AS AN APPLIED SCIENCE

The rapid advances in technology of the past decade have enriched our reserve of mineral resources while reducing costs and have aided in man's quest to learn about the oceans.

Desalination processes have improved substantially. Today, over 100 million gallons of processed water are used daily (Fig. 11-14). New methods of desalination have been employed in an attempt to reduce the cost of preparation. Although most of the desalted water is prepared by distillation processes, experiments have been conducted using alternate methods such as vacuum freezing and reverse osmosis. The cost of preparing desalted water is still too high to allow its use for irrigation; for the most part the water is being used for household purposes, but with a less expensive method of desalination the list of possible uses will be limitless.

With the advances being made in offshore drilling techniques the

11-14   A desalting plant in Mauritania. *(FAO-UN)*

offshore oil production has moved to deeper and rougher waters as well as toward the more remote places of the ocean floor (Fig. 11-15). Once again, the major drawback in utilizing our newly found untapped source of oil is the cost. To be profitable the cost of drilling deep-sea oil wells should not exceed the costs of drilling in 200 feet of water. As more improvements are made in the deep-sea drilling system, the cost of obtaining oil from the ocean floor will be greatly reduced.

Until recently, the drilling sites were limited to the shallower parts of the ocean because of the necessity of using stationary drilling platforms on which to work. It is now possible to use an unanchored deep-ocean drilling system. The *Glomar Challenger,* operated by Scripps Institution of Oceanography, has been successfully drilling in water depths greater than 17,000 feet.

The development of fiber glass and plastics has also enhanced the future of oceanography. These strong and versatile products—often as strong as the metals they replace—are immune to the corrosive effects of the sea. This alone saves money by reducing the percentage of equipment failure caused by contact with the seawater.

The construction of large supertankers, capable of carrying several hundred thousand tons of cargo, has reduced shipping costs. These

11-15 An offshore drilling platform in the North Sea 140 miles off the English coast. [*Standard Oil Co. (N.J.)*]

---

costs had been mounting rapidly with the use of smaller ships. Other new cargo vessels, which often exceed 1,000 feet in length, carry cargo sealed in tamperproof containers. Containerization has also reduced the cost of moving goods to the shipper. The containers are loaded at the exporter's warehouse. The contents are sealed at the packing point, thus reducing anticipated losses due to pilferage and damage from the elements. The containers are mounted on trucks and driven to the shipping pier. Since the containers are for the most part a standard size, the time for loading is greatly reduced. This results in a reduced labor cost. The sealed containers are shipped to their port on flatbed trains to be shipped to the inland customer. Again the amount of handling of the individual produce has been reduced, resulting in further savings for the wholesaler, which may in turn be passed on to the retail customers.

The increase in oceanographic activity has led to a worldwide system of cooperative data sharing. Although most oceanic research is being carried on by only a few major countries, Canada, France, Germany, Japan, the United Kingdom, the United States, and the

Soviet Union, many other countries use the oceans as a source of food and natural resources.

The International Geophysical Year, established during 1957–1958 by the International Union of Geodesy and Geophysics, created a series of worldwide information centers some of which are still in operation today. These centers provide a clearinghouse for the international exchange of oceanographic data. In recent years governments have also established commissions with the expressed purpose of sharing data collected by oceanographic investigators from all parts of the world. UNESCO, the United Nations Educational, Scientific, and Cultural Organization, has been a leader in this field for over the past decade.

In addition to the international governmental sharing of data, which is a necessity to assimilate the proliferation of data, a cooperative attitude exists among the research teams of the various laboratories conducting deep-sea research. The spirit of cooperation has provided a number of services not only important to research oceanographers but to other mariners as well.

Bathymetric charts of the continental margins are coordinated by the International Hydrographic Organization in Monaco, which provides this service to mariners. In the past, the sailor's concern was for the coastal waters, where he might run aground, but the future will require a similar series of charts to provide topographic information for the deep-sea floor as well. Scientists, fishermen, oil companies, and a number of governmental agencies will use these data to expand their exploitation and knowledge of the sea.

Weather forecasting has been improved through the use of a network of ships and island stations to report observations to a central location for analysis and dissemination of the data. Although the number of stations required for accurate forecasting is far from complete, a reasonably reliable forecast can be provided on a day-to-day basis. The deficiencies in the network will be filled in the next few years with the addition of more stations and greater use of orbiting satellites to monitor and predict the long-range weather on any part of the sea. This World Weather Watch is being coordinated by the World Meteorological Organization, an agency of the United Nations.

International cooperation has broken down in the legal and political determination of boundaries of the various countries. Oceanographic phenomena do not respect boundary lines drawn on charts, and the research must be conducted where the phenomena exist. However, a number of countries have objected strenuously not only to fishing vessels which have invaded their territorial waters but even to the presence of research ships in these waters (Fig. 11-16). Since there are no established guidelines to determine the territorial extent of the continental boundaries in the sea, the stated boundary varies from country to country. As the seacoast countries realize the potential of their nearshore waters, they are becoming more and more reluctant to allow foreign research vessels to investigate them. The United Nations, recognizing the importance of international cooperation in the use of the ocean and its resources, has established a study commission to develop recommendations in regard to the peaceful uses of the sea.

A similar problem exists in attempting to establish guidelines for international fisheries in the oceans. The 1958 Geneva Convention essentially gave sovereign rights to the continental shelf to the coastal state out to a water depth of 200 meters or to where the technology can reach. It is this exploitability clause that has caused much of the problem in determining the legal extent of the sovereign do-

11-16   Commercial fishing in the Pacific. *(FAO-UN)*

main. As the technology of a country increases, they are able to claim more and more of the seabed, thus reducing the use of the sea by other countries. To ensure the preservation of the sea for future use — both as a source of natural resources and productive fishery — stronger regulatory machinery is urgently needed.

SUMMARY QUESTIONS

1. What is rapture of the deep? How does it severely limit the depth to which a diver may descend?

2. What is a bathyscaphe? How may it be compared to another type of craft used by man in another sphere of the planet?

3. Why are programs like Sealab and Tektite being conducted since they are run at relatively shallow depths? What data are they specifically designed to reveal?

4. How are meteorological studies improved through ocean stations and cooperative networks?

5. List some of the major technological advances responsible for man's exploration of the sea.

6. How does today's diving apparatus differ from the earlier equipment used by divers?

7. Why is it necessary for man to decompress during ascent from deep or extended dives in the sea?

8. What are some of the problems in the economic use of desalted water?

9. How has the development of large supertankers revolutionized the shipping industry?

10. What are the components of SCUBA apparatus? How do they operate?

11. How do the conditions in a submersible differ from an underwater habitat?

12. Describe the operation of the *Trieste.*

13. What does bathyscaphe mean? Mesoscaphe?

14. What is containerization?

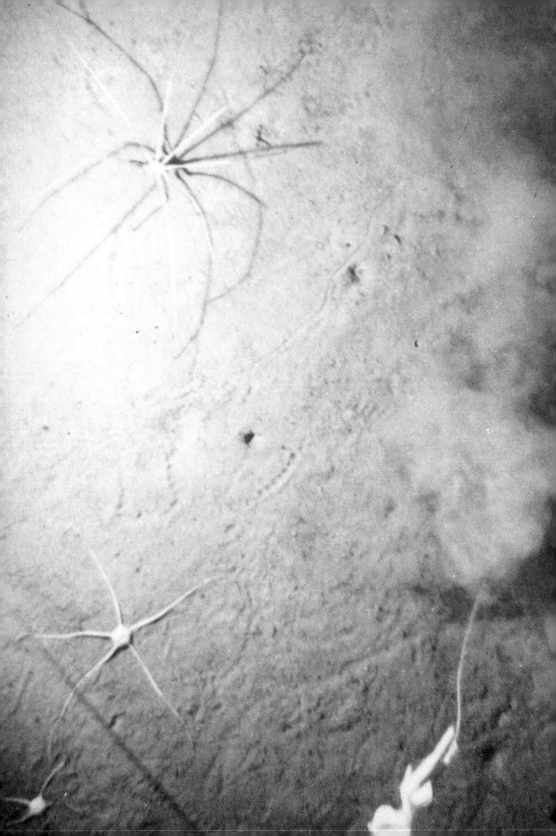

12 | LIFE IN THE OCEAN

Even casual observation of the sea reveals
that it supports a diverse and abundant variety of life-forms. However, the specialized
evolutionary changes that fit marine organisms for their habitats are not so obvious.
The distribution and density of sea life depend
on many physical and chemical properties of
the marine environment, the most important
being the variable conditions of light, temperature, pressure, and salinity. Generally, the
basic requirements for life on land are also
necessary to support life in the oceans, that is,
food, shelter, protection, and waste elimina-
tion. Marine life is directly affected by critical
changes in the materials necessary to build
protoplasm and provide energy to the organism. For this reason, the physical characteristics of the water, sunlight, heat, oxygen, and
carbon dioxide greatly affect the development of life in the sea. The type of life in the
sea varies with the changing conditions (see
Fig. 12-1). Life on Earth, which probably started in the ancient seas more than 3 billion
years ago, has evolved into a complexity of
forms of plant and animal types. The flora
and fauna of the oceans exist in an ecological
relationship in which plants serve as a source
of food supply for some of the animals, which
in turn are a food source for other animals,
while the animals return to the sea the nutrient elements which decompose and become
available to the plants.

THE PHYSICAL AND CHEMICAL FACTORS OF SEA LIFE

Although the physical and chemical factors of
the marine environment are considered in-

Deep sea sediment and sea floor life. (Photograph published with the permission of the Woods Hole Oceanographic Institution)

Eras and duration (millions of years)	Periods (millions of years from present)		Event
Cenozoic (65)	Quaternary (0–2.5)	Recent Pleistocene	Man ↑ Four glacial periods
	Tertiary (2.5–65)	Pliocene Miocene Oligocene Eocene Paleocene	Mammals
Mesozoic (160)	Late — Cretaceous (65–136)		Flowering plants become dominant
	Middle — Jurassic (136–190)		First birds—flowering plants appear
	Early — Triassic (190–225)		Dinosaur period
Paleozoic (345)	Late — Permian (225–280)		First reptiles
	Carboniferous (280–345)		Fossil fuels origin from fern forests
	Middle — Devonian (345–395)		Amphibians appear
	Silurian (395–430)		Marine algae—Common insects arise
	Early — Ordovician (430–500)		Land plants arise—Fish fossil appears
	Cambrian (500–570)		75% of North America a shallow sea
Proterozoic (930)	Precambrian (570–1500)		Algae appear
Archeozoic (3000)	(1500–4500)		Approximate age of oldest rock No fossils found

12-1   Evolutionary history of plants and animals in the sea.

dividually here, they are in fact totally interdependent, acting in conjunction with one another to produce the conditions necessary to support life in the sea.

THE PHYSICAL FACTORS

The physical characteristics of the ocean have been of great interest to scientists. Today, the marine scientist uses highly sensitive instruments to study the worldwide characteristics of the Earth's oceans. As in the past, the modern oceanographer investigates the physical properties of pressure, temperature, viscosity, density, circulation, and light to aid him in developing a better understanding of the complex pattern of the sea.

Pressure

The effects of pressure do not exclude life in the abyssal depths of the sea. The pressure in the ocean increases by 1 atmosphere for every 33 feet of depth. Pressure reaches a maximum of approximately 1,000 atmospheres at its greatest depth. However, fish are so adapted that their swim bladder adjusts to compensate for external changes in pressure. As the external pressure increases, the swim bladder is compressed, and similarly, as the pressure is reduced, the swim bladder expands. By controlling the amount of gas in the bladder, the specific gravity of the fish is kept at the same level as the surrounding water.

Nevertheless, changes in pressure often serve as a barrier to the vertical movement of animals. Gases are more soluble in water at high pressure than at low pressure; however, atmospheric gases are not readily available in deep water. Thus, even though pressures are extremely high in deep water, dissolved gases are lacking and less dissolved oxygen is available to deep-water fish.

Pressure also has an effect on the temperature of water. The temperature of deep water is higher than it would be if the pressure were less. Theoretically, these slight changes in temperature should have some effect on the type of organism living in the ocean, but the actual effects are unknown at this time.

Temperature

Although the relationship of slight variations in water temperature to life-forms is uncertain, temperature largely controls the rate at which chemical reactions take place. A change in temperature would affect the chemical processes of life, or metabolism, continuously taking place in plant and animal tissue and would thereby affect the organism as a whole. In warm water, the rate of photosynthesis in plants is increased. Certain species of animals grow faster in warm water. The organism that grows faster as a result of higher temperature usually attains a smaller adult size. The same species living in cold water grows more slowly but attains a larger size and even lives longer (Fig. 12-2).

Carbon dioxide, a compound essential to plants, is more soluble in cold water than in warm water. This accounts for the unexpected luxuriant development of plant life in the colder waters.

By means of the *bathythermograph,* an instrument for measuring the temperature of various levels of the sea, a profile of temperature stratification has been drawn. Generally, temperature gradients fall into three categories based on zonal lines of constant temperature called *isotherms.* The first level, a *mixed* layer, is located close to the

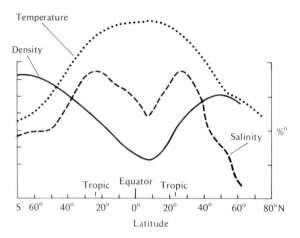

12-2   Variations with latitude of surface temperature, salinity, and density.

surface and extends to a depth of 50 to 200 meters. This layer maintains an average temperature similar to that of the surface. In the second layer, an intermediate zone extending to a depth of 500 to 1,000 meters, there is a rapid decrease in temperature. In the deep layer, or third zone, temperature changes take place at a very slow rate.

A typical temperature gradient in the low latitudes shows a temperature of about 20 °C at the surface, 8 °C at the 500-meter level, 5 °C at the middle-layer 1,000-meter level, and about 2 °C at a depth of 4,000 meters.

Ocean temperatures are also indirectly responsible for other physical properties of the marine environment. Temperature determines the amount of dissolved gases in solution, the viscosity of the water, and density; all three of these affect the distribution of life in the seas.

Viscosity

Water, like all liquids, is subject to change in its ability to flow. This property, called *viscosity,* is an internal resistance resulting when particles of the fluid adhere to each other. The adhesion develops a force which opposes the flow. Under a set of fixed environmental conditions, fluids naturally differ in their ability to flow. For example, alcohol always flows more easily than molasses. Changes in environmental factors change the viscosity of a particular fluid. Temperature has an effect on the rate of flow. Molasses flows more readily at room temperature than if it has just been taken out of the refrigerator.

The viscosity of all liquids decreases with an increase in temperature. A warmer liquid is less viscous and therefore flows more rapidly than a colder sample of the same liquid.

The importance of viscosity to the distribution of marine life is often overlooked. Viscosity may impede the movement of organisms through the water. An animal must expend larger amounts of energy to swim the same distance through cold water than through warm water. The natural streamlined form of fish has helped overcome the resistance to movement produced by viscosity. To some extent viscosity has established patterns of distribution for sea life. Generally, the upper waters are warm and less viscous while the deeper waters are cool and more viscous. Thus, the adults of particular species tend to live more deeply than the younger and smaller of that same species, which have greater surface area in proportion to bulk.

Even within a single species, adult forms living under different conditions reveal adaptations to the environment. For example, a group of dinoflagellates, the *Ceratium,* have hornlike projections the length of which is governed by the viscosity of the water. In warm water species, the horns grow longer and help prevent them from sinking due to the reduced viscosity. In colder water species, the horns are considerably shorter, as an adaptation to the increased viscosity.

Density
Density, or the mass of a unit volume of water, depends upon temperature and salinity. The higher the temperature the lower the density for a given salinity; and the higher the salinity the higher the density for a given temperature. In the ocean, water of a high density usually sinks below water that is less dense. This is the fundamental cause of the continual circulation of ocean waters.

Circulation
The multitude of marine life now in the oceans would not be possible if the water of the oceans were not in constant circulation. The rhythmic and progressive movement of the water distributes the nutrient materials necessary for plant and animal growth in the sea. It also helps in the diffusion of important gases needed to support marine life. The constant motion of the water also serves as a transporting medium for many life-forms which cannot provide their own locomotion. Water currents are produced by the force of the prevailing winds blowing over the ocean surface and by differences in density of the seawater. The wide dispersal of marine life is the most obvious outcome of the movement of the sea.

Light

Light, extremely important to the development of plant life in the sea, affects animal life indirectly because it depends upon plants for food. Light provides the energy for photosynthesis necessary to maintain the green plant. For this reason, the depth to which sunlight penetrates the ocean determines the vertical distribution of green plants. The degree of penetration is based on the amount of solar energy striking the surface of the ocean (Fig. 12-3). The infrared, or heat, portion of the electromagnetic spectrum is quickly absorbed and does not penetrate deeper than 1 meter. The visible portion of the spectrum, which makes up a larger percentage of the solar energy available at the surface, can penetrate deeper. This portion of the spectrum may also be converted to heat, and thus be absorbed by the water. The quantity of suspended particles in seawater also decreases the depth of light penetration. However, even in particle-free water, light can penetrate only the surface zone. Almost all visible light has been absorbed by the 100-meter depth.

The apparent color of the ocean water is an indication of the amount of suspended material it carries. For example, blue-green rays penetrate deepest; therefore the blue-green color of the ocean surface indicates that this water is relatively free from suspended particles. A concentration of suspended particles changes the amount of scattering and produces a greenish color at the water surface. Depending upon the concentration of particles, the water may take on a brownish or reddish hue (but red is usually caused by the growth of red algae).

Aside from their dependence on plants for nutrition, marine animals can exist without sunlight. The absence of light stimulates changes in the body functions and pigmentation of the marine fauna. Animal life below the level of light penetration is often darkly pigmented. In most cases, animals with little vertical migration are blind. Some lack eyes, which probably atrophied in ancestral forms from

---

12-3 The electromagnetic spectrum.

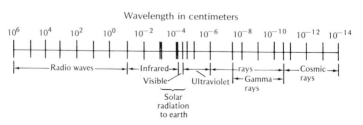

lack of use. Some animals of the deep sea produce their own light. The purpose of this phenomenon, known as *bioluminescence,* is not known, but it may aid deep-sea creatures with poorly developed eyes. Some bioluminescence probably helps attract prey; in other cases it helps find mates of the same species. At the other extreme, some organisms have developed enlarged eyes and even telescopic eyes that protrude from the eye sockets. Thus, we can find adaptations in both directions (Fig. 12-4).

Planktonic animals, animal forms which drift with the moving waters, also respond to changes in available light. Many varieties of planktonic animals move closer to the surface at night and sink deeper during the day. This vertical migration pattern continues to be an unsolved puzzle, but the changing light intensity is the primary cause for movements. The diurnal variations in depth level of these animals produces a *deep scattering* layer, or false bottom (Fig. 12-5), which is recorded by echo soundings at a higher level during the night than during the day. Marine ecologists hypothesize that the planktonic animals' ability to avoid light has adaptive significance. The animals are able to avoid predators in the dark. Living in this dark zone, these organisms also avoid harmful side effects produced by some plant toxins released by certain green plants located in the photic zone.

THE CHEMICAL FACTORS

Seawater is a complex chemical mixture, and even to begin to understand it we must look at interactions between individual factors. In the future, advanced technology will enable man to remove many valuable minerals from seawater. In this book our investigation is limited to the types of chemical reactions that occur and the effects they have on the plant and animal life in the sea. Among the impor-

---

12-4   Deep-sea fish.

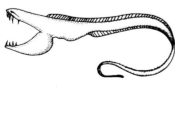

(a)   (b)

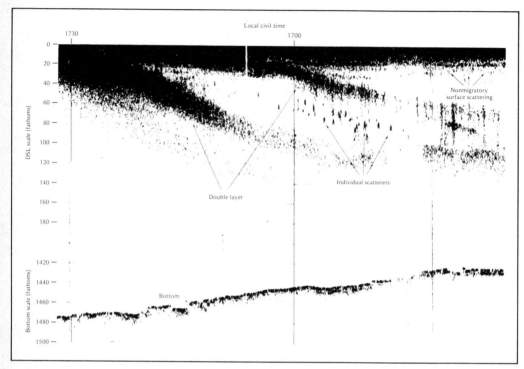

12-5   An echogram showing the deep scattering layer. *(USNOO)*

tant chemical factors are salinity, oxygen content, carbon dioxide concentration, and the role of inorganic elements in determining the chemical properties of seawater (see Table 12-1).

Salinity

The measure of the amount of salt dissolved in the ocean water is called *salinity*. Expressed in terms of so many parts per thousand parts of water, the average distribution is about 35 parts per thousand. Although the marine organisms are able, for the most part, to maintain a balance between the seawater and their body fluids, salinity is a contributing factor to the distribution of life in the seas. According to their ability to tolerate specific concentrations of sea salts, organisms are classified as *stenohaline* or *euryhaline*.

Stenohaline organisms are sensitive to small changes in salinity. Characteristic of deep-sea forms, these organisms may die if driven into coastal waters by wind or currents. Most marine organisms fall into this category. The euryhaline forms, on the other hand, have high tolerance for changes in the concentration of dissolved salts in

TABLE 12-1 The Major Ionic Constituents Dissolved in Seawater

Ion	Name	Percentage of all dissolved material
Cl^-	Chloride	55.04
Na^+	Sodium	30.61
SO_4^{--}	Sulfate	7.68
Mg^{++}	Magnesium	3.69
Ca^{++}	Calcium	1.16
K^+	Potassium	1.10
HCO_3^-	Bicarbonate	.41
		99.69

the water and thus can pass from salt to freshwater at will. For example, the salmon, the shad, and the eel live part of their life in salt water and part in freshwater. Salinity does not affect the rate of reproduction, but it controls what type of animal can live in a certain environment (see Fig. 12-6).

In general, the maximum size of the euryhaline type is smaller than that of the stenohaline type, although among members of the same species of euryhaline life-forms those living in waters of lower salinity will reach a smaller maximum size.

12-6 A school of shallow-water marine fish. *(USNOO)*

Oxygen

Life in the sea requires dissolved oxygen in the seawater. The life processes of all organisms depend upon oxygen to carry on their normal metabolic functions. Animals in the sea consume oxygen, which must be replaced by the photosynthetic plants. A few forms, such as anaerobic bacteria, can carry on respiration by releasing oxygen from organic compounds. Other types, such as the mussel and clam, can survive for a short time in water of low oxygen content. During this time the organisms become inactive. Although it is rare for ocean water to be permanently devoid of oxygen, if the condition persists, the life-forms soon die. When this situation does occur, the area soon becomes devoid of animal life. The ocean sediments blacken and are frequently accompanied by the smell of hydrogen sulfide gas, indicating decay of the organisms. The seawater absorbs oxygen when it is exposed to the surface, and the concentration of hydrogen sulfide slowly drops as the water mixes and circulates. The amount of oxygen in the ocean is much less than the available oxygen in the atmosphere, the values being 9 and 200 milliliters per liter, respectively. Oxygen and carbon dioxide are the two most important dissolved gases in the sea.

Carbon Dioxide

Carbon dioxide is necessary for the photosynthetic process of plants and is produced as a waste product of respiration by both plants and animals (Fig. 12-7). The sea contains a much greater volume of dissolved carbon dioxide than the atmosphere—60 times as much. The sea in fact serves to regulate the gas when the atmospheric level of carbon dioxide becomes excessive. The carbon dioxide of the sea is used directly by marine plants. The carbon compounds of all marine organisms—from protozoa to the higher forms of animals—come from the carbon dioxide in the sea.

Carbon dioxide is also ecologically important in the deposition and solution of calcium carbonate, $CaCO_3$. The precipitation of calcium occurs most readily under conditions of high temperature and salinity and low carbon dioxide, CO_2, concentration. Calcium is a necessary element for the formation of skeletons and shells in mollusks, crustaceans, and other life-forms. In the deep ocean, water temperature is low and carbon dioxide concentration is high due to the lack of photosynthetic plants. Therefore precipitation of calcium does not occur readily in deep water, and most life-forms which require calcium deposits are limited to relatively shallow waters.

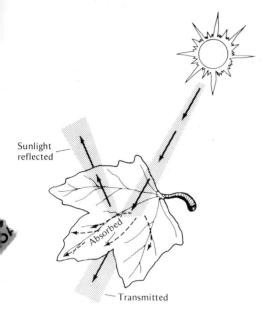

Sunlight reflected

Absorbed

Transmitted

12-7   Sunlight interacts with green plants, resulting in photosynthesis.

Inorganic Elements

A chemical analysis of seawater reveals that most of the elements necessary for the support of plant and animal life are abundant. Some of the elements appear as traces so slight that they are hardly recognized by ordinary means, but even these trace elements play a significant part in the ecology of the sea. Inorganic compounds of the sea contain abundant quantities of carbon, oxygen, nitrogen, and phosphorus.

As on land, the primary biological activity of green plants is the process of photosynthesis. Carbon, nitrogen, and phosphorus are removed from the water solution when sufficient light penetrates for photosynthesis, usually in the top 100 meters. This results in a depletion of nutrient materials from the surface waters, leaving an enriched concentration in the deeper waters. If it were not for circulation, biological activity would not continue. The trace elements occur in very low concentrations and are probably kept in balance ecologically by the action of marine organisms. Plankton, for example, tend to remove many of the trace elements from the sea. Such trace elements as copper, zinc, cobalt, iron, and molybdenum are known to be important to biological systems.

THE MARINE ENVIRONMENT

The modes of living and habitats, from which marine organisms do not stray, vary greatly. The marine environment is divided into distinct realms based on the characteristic ecological features and the plants and animals associated with them. To facilitate the study of marine life the sea has been divided into zones according to its physical and chemical properties. The boundaries, which often overlap, require subdivisions to account for recognizable changes in the life-forms present in each environment. The two main divisions of the sea are the *benthic* and the *pelagic* zones (Fig. 12-8), respectively the floor of the ocean and the water mass itself.

THE BENTHIC ENVIRONMENT

The benthic, or bottom, terrain of the ocean stretches from the tide level to the greatest ocean depths. It supports the bottom-dwelling life of the ocean. The term *benthos* is used to describe all organisms that live on or in the bottom of the oceans. The benthic realm is subdivided into the *littoral* and *deep-sea zones*. The outer limits of the continental shelf mark the edge of the littoral zone, which is further subdivided into the *intertidal, eulittoral,* and the *sublittoral* zones. The deep-sea zone is subdivided into an upper, *bathyal,* zone and a lower, *abyssal,* zone.

For the most part, the delineation of these zones is based on ocean depth and the life-form distribution they support. The intertidal zone is located above the low-water mark; the eulittoral zone reaches from the low-water line to the point where the amount of

12-8   (a) The division of the marine environment.

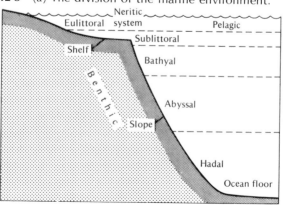

(a)

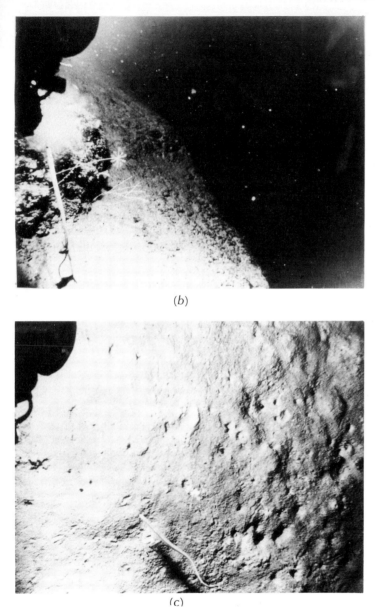

(b)

(c)

12-8   (b) Pelagic life-forms.
(c) Benthic life-forms of the deep sea. (*Photograph published with the permission of the Woods Hole Oceanographic Institution*)

light penetration will support the bottom-dwelling aquatic plants; and the sublittoral zone extends to the limits of the continental shelf, a depth of approximately 200 meters. The deep sea exists beyond the littoral zone, where there is no effective light penetration. This zone consists of the bathyal zone of the continental slope and the abyssal zone, which begins beyond a depth of 1,000 meters.

Benthic Life-Forms

The bottom-dwelling creatures of the ocean make up the benthonic life-forms, which are both sessile and motile. These organisms play an important part in the food chain of sea life and are distributed from the nearshore littoral regions to the deep-sea bottom. The zonation of the organisms has an effect on the development of the specific life-forms. For example, the animals of the littoral benthic zone are adapted to seasonal and daily changes in temperature and salinity. Life in the shallower areas of the littoral zone must be able to withstand the action of waves and tides.

Benthic animals have developed various adaptations to the type of bottom material, depth, and feeding habit. Benthic life-forms have three main methods of feeding; some live on the debris produced by the breakdown of plant and animal bodies which is constantly settling to the ocean bottom, others filter the suspended organic particles directly from water, and still others depend on their predatory abilities to secure food in the form of bottom-dwelling fish. The sessile forms, which must anchor themselves to the sea floor to combat the turbulent conditions of waves and tides, depend for survival upon the types of material to which they must be attached. They cannot go and search for food but must rely on what floats within their reach.

Some shelled organisms and sea worms are able to burrow into the soft mud; oysters and barnacles can cement themselves to firm surfaces such as rocks and pilings; and other forms must be able to withstand periodic intervals without any water at all. Animals living in the intertidal zone have developed adaptations that allow them to withstand long periods of direct sunlight without water. In certain regions, where at times the changes are more enduring and whole populations are subject to destruction, only the most tolerant will survive.

The deep-sea benthic life, existing below a depth of 300 meters, is mostly mud-dwelling types which have adapted to this form of life. In the absence of sunlight green plants do not grow, and thus the benthic animals must depend for nutrition upon the organisms which filter down from above. The deep-sea fauna show adaptations consistent with the continual darkness and slow movement of the ocean depths. The animal population of this deep-sea zone decreases significantly with depth and distance from the shore. The limited food supply means that even the number of individuals of any life-form is not uniform, being dependent upon the amount of available food settling from above.

THE PELAGIC ENVIRONMENT

The pelagic region of the oceans consists of all the water covering the benthic environment. It may be subdivided horizontally as well as vertically. The horizontal subdivision is composed of the *neritic,* or *inshore,* province and the *oceanic,* deep-sea province. Vertically, the pelagic realm is divided into an upper, *euphotic,* zone, which marks the limit of photosynthesis and light penetration, and lower, *aphotic,* zone beyond a depth of 200 meters, to which no light penetrates.

Planktonic Life-Forms

Plankton is a word from the Greek and means wandering or drifting. Plankton makes up the free-floating or drifting organisms of the sea. For the most part these organisms are incapable of self-locomotion and are carried by ocean currents. Plankton is divided into two groups, *phytoplankton* and *zooplankton.* The phytoplankton consists of the myriad of floating plants such as diatoms, dinoflagellates, coccolithopores, and sargasso weeds. The zooplankton, or floating animal life of the sea, relies upon the phytoplankton as a source of food. Typical zooplankton includes countless animal eggs and larvae, tiny jellyfish, worms, small crustaceans, and copepods suspended in the water (Fig. 12-9).

Nektonic Life-Forms

Nekton refers to marine organisms capable of self-locomotion. Unlike the plankton, which must drift from place to place in the open sea, nektonic life-forms can swim long distances for extended periods of time. Their locomotive abilities not only aid them in selecting a suitable environment in which to live but also help them seek protection from a natural enemy when under attack. While most fish are nektonic in nature, their eggs or larvae are often planktonic. The young, weak-swimming fry continue to drift with the ocean current. The adult fish often have migratory instincts which make them return to the place where they were spawned. The nektonic life-forms are

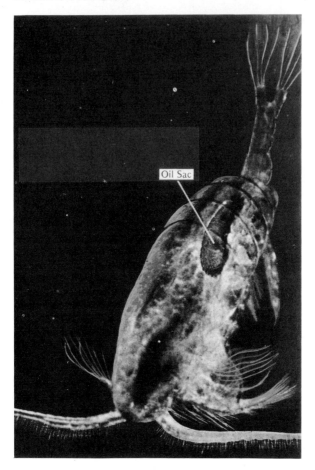

12-9 Enlarged photograph of the 1/8-inch marine copepod *Calanus,* known as the "insect of the sea." The animal's oil sac is plainly indicated toward the rear of its tiny body. Wax is used for energy storage and as a food supply during periods of starvation and hibernation in long winters. It is made by the animal from oils of algae that it eats (darker area below sac). *(Scripps Institution of Oceanography, University of California, San Diego)*

directly affected by physical barriers which tend to keep a particular species within clearly defined limits in the open sea. Pressure, temperature, density, and nutritive condition; chemical composition, including dissolved gases, salinity, and other inorganic elements; and circulatory patterns—all these work together to keep the diverse life in the sea separated in specific geographical ranges.

SUMMARY QUESTIONS

1. How are pressure and gas solubility related? Temperature and solubility?

2. How do the young of a species adapt to the problem of viscosity?

3. Describe the various factors that may affect the color of water in the open sea.

4. If life-forms decline in number with depth, how are deep-sea organisms adapted to survival in such harsh conditions?

5. What are stenohaline and euryhaline organisms? How do they differ?

6. Describe the layers of the benthic environment and the general nature of life in each.

7. What is plankton and what is its major importance in the food chain?

8. What are the advantages of organisms capable of vertical migration?

9. Describe the classification scheme of biological marine populations with respect to locomotive ability.

10. Describe the generalized temperature stratification of the open sea.

11. How does viscosity serve to limit the vertical migration of organisms in the sea?

12. What is a bathythermograph?

13. What is the mixed layer in the ocean?

14. What produces the usual blue-green color of the open sea?

15. What is the deep scattering layer?

16. What is the major life form in the deep-sea benthic zone?

13 | FOOD RESOURCES FROM THE SEA

Although it is not possible to determine when man first made use of the food resources of the ocean, the discovery of ancient fishing implements makes it possible to state that man resorted to the sea as a source of food several thousand years before the birth of Christ. Whether the oceans first served our ancestors as a mode of transportation or as a source of food will probably never be decided. However, present and future uses for the resources of the ocean reflect a multiplicity of purpose. We may hope for new and abundant sources of food and minerals.

As the critical problem of protein deficiency increases with the growing population of the world, man's ability to produce adequate food supplies is diminishing and his dependency on the oceans is becoming more evident. Although the current annual yield from world fisheries exceeds 100 billion pounds, this can be increased by more intensive development of the fisheries. Man has a daily requirement of 10 to 20 grams of protein per person, which means an annual protein intake of between 8 and 16 pounds. Since it is clear that there is a definite need for a protein-rich diet and the ocean has an abundant supply of protein, a means of obtaining that protein with the greatest efficiency is obviously desirable.

FOOD CHAIN IN THE SEA

The present harvest of edible fish from the oceans represents the end products of a food chain (Fig. 13-1) that begins with phytoplankton, the primary food source of the sea. How-

A midwater trawl. *(Scripps Institution of Oceanography)*

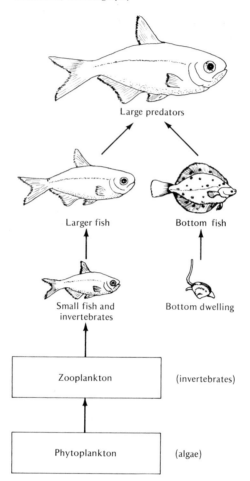

Large predators

Larger fish   **Bottom fish**

Small fish and   **Bottom dwelling**
invertebrates

| Zooplankton | (invertebrates) |

| Phytoplankton | (algae) |

13-1   A generalized scheme of the food chain in the sea.

ever, the concentration of plant material in the sea is such that it would require a million pounds of water to yield 1 pound of plant material. Although the phytoplankton is the most efficient concentrator of nutrient materials, the inefficiency of collecting and the lack of appetizing appeal of this plant material have led man to search higher up the food chain for nutrients. One difficulty with this is the reduction in available nutrient material. There is apparently a 10 percent conversion factor in each of the various trophic levels. For example, if man depends on organisms that are two food levels above the phytoplankton, he is receiving the equivalent of 1 percent of the nutrient value of the original amount of plankton consumed. In the first level only 10 percent of the available carbon would be converted to organic materials, while in the second level 10 percent of the 10 percent, or 1 percent, of the original quantity would be converted.

A typical food chain begins with the phytoplankton eaten by herbivores, small plant-eating animals which strain the phytoplankton from the water (Fig. 13-2). The plant material is then converted into animal protein. The herbivores in turn are eaten by carnivores, which themselves fall prey to the larger sea organisms. This cycle is continuous, but each step above the original source results in a loss of energy as part of the food consumed is burned to provide energy for the life processes and energy expended in capturing the food. Thus there is a definite advantage to using organisms farther down in the food cycle. Although some of the marine products harvested from the sea are herbivores, the greatest percentage of the harvest is derived from some higher trophic level. Since the annual ocean yield is insufficient for man's purposes, it would be advantageous to alter our habits and increase the utilization of animals from the lower trophic levels (Fig. 13-3).

13-2 *Discocoasters,* an ancient planktonic life-form. (*Lamont-Doherty Geological Observatory*)

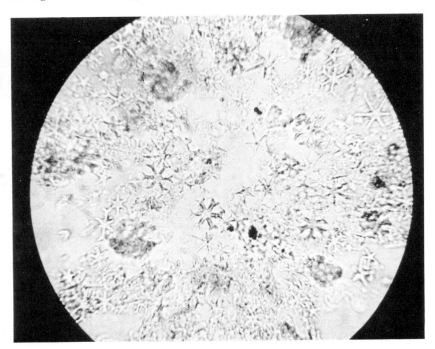

(a)

(b)

13-3 *(a)* Plankton collecting. *(Lamont-Doherty Geological Observatory)*
*(b) A plankton net. (Photograph published with the permission of Woods
Hole Oceanographic Institution)*

AQUACULTURE

Recognizing that the fish population in the world's oceans will be a critical factor in the future, some fishermen have taken to raising marine organisms under controlled conditions rather than catching them. This practice, called *aquaculture,* is similar to farming on land. At present, this process usually is limited to clams, oysters, and shrimp. But just as the freshwaters of the world have been stocked with fish for many years, it is possible to hatch and develop fish for the sea. Although aquaculture is now limited to estuaries and coastal waters, it is possible to manage the yield much as we manage our farmlands.

The science of aquaculture as practiced outside the United States has resulted in an extremely efficient production of marine protein, more efficient than if nature were allowed to take its course. A major problem facing these "fish farmers" is the potential pollution of the waters being used for aquaculture. We use the same waters in many cases for the disposal of wastes, which results in the development of undesirable and inedible species of marine algae which have replaced the natural populations of some areas. This has limited and even prevented the growth of marine organisms in some instances.

In cultivating shellfish man has used natural processes to increase the productivity of these organisms. The shellfish are often suspended on a vertical wire, sometimes to a depth of 30 meters. This method of farming results in the increase of available water volume, even though the actual surface area covered is limited.

The nonmobile forms of shellfish have an additional advantage over the mobile marine organisms. The phytoplankton upon which they feed is carried to them by the tidal currents. The conversion of food into animal flesh is more efficient than it would be in an organism that must actively pursue its food. Through the process of aquaculture the shellfish are allowed to develop in a protected situation, eliminating the problem of natural predators such as starfish. The suspension of the shellfish on vertical strands also increases the amount of available food, as the bottom-dwelling forms must depend on the plankton that settles to the bottom while the suspended forms filter the plankton directly from the water.

Besides the technological skills necessary to carry on a viable program of aquaculture, a strong initiative and high incentive are necessary. The United States has not been very active in the field of aquaculture, due less to a lack of scientific know-how than to the lack of incentive. Other countries, such as Japan and Spain, are presently taking full advantage of their nearshore waters to utilize their full capacity for food production (Fig. 13-4).

(a)

(b)

(c)

13-4   (a) Oyster farms in Japan.
(b) Preparing the suspension of oysters in protective baskets.
(c) Harvesting the oyster crop. (*Consulate General of Japan, N.Y.*)

WHO OWNS THE SEAS?

One of the major issues slowing the progress and development of aquaculture is the legal aspect of farming the nearshore waters. By and large the coastal waters are considered public domain, and any attempt to restrict the use of specific areas is usually met by public opposition.

In addition to legal constraints a number of technological problems have slowed the development of aquaculture. Attempts to hatch large quantities of shellfish under controlled conditions have met with setbacks. The mortality rate of developing shellfish is high compared with that experienced in natural reproduction, particularly in countries that have not adopted the Japanese three-dimensional method of farming the shellfish. Where bottom beds are used, the shellfish are subject to the effects of natural predators as well as the possibility of suffocation by the moving sediments of the ocean bottom.

THE FUTURE OF AQUACULTURE

If the science of aquaculture is to become a profitable and viable industry, a method to reduce production costs must be found. The development of artificial foods is a necessity. At the present time, it is necessary to develop one species for food in order to market another species. Experiments are currently being conducted to develop methods of raising large quantities of algae, the primary producer, at a cost which will not be prohibitive. Since cultivated algae have been fed to livestock with successful results, it is hoped that they can someday be fed to the marine organisms being raised by aquaculture.

Selective breeding to produce offspring with superior qualities is suggested for some species. Experiments to control physical and chemical characteristics of water, such as temperature and salinity, are also being considered.

Several experiments with aquaculture of fish reveal the possible success of this venture. In Taiwan, large open coastal areas have been converted to fish farms in order to develop food stocks under controlled situations. This practice of raising fish in a small area yields about 700 tons of fish per square mile each year. The catch is enhanced by the addition of nutrients to the water in order to increase the efficiency of the farm. In Indonesia, sewage is used to fertilize the farm. Although this method has esthetic and hygienic disadvantages, its success is unmistakable. In this case, the yield is 1,300 tons of fish per square mile.

In both cases, the results can be compared with those in the

Philippines, where fish farming yields about 300 tons of fish per square mile. Although no fertilizers or replacement nutrients are used, the yield is far greater than that obtained in an average sea catch (70 tons per square mile). These results suggest that success can be achieved by using treated sewage as a fertilizing medium for various marine food stocks. Experiments such as these are currently being carried out in the United States.

Although many advances are being made to increase the efficiency of the yield through aquaculture, it is foolhardy to assume that this will answer all the world's nutritional problems. Presently less than 5 percent of the world's food supply is derived from the sea, and even if this yield were doubled in the next few years, it would do little to reduce the problem of malnutrition existing in the many underdeveloped countries which contain about two-thirds of the world's population. Aquaculture will not erase hunger, but it will surely forestall famine in many parts of the world. Aquaculture represents a program of efficient use and management of one of our most prized natural resources.

One thing that must be considered in attempting to increase the yield of food fish from the sea is the production rate of phytoplankton. An increase in the production of phytoplankton would be desirable, but in order to achieve such an increase in the sea the natural fertility of the water must be increased. It is possible to add fertilizer directly to the sea, but, with the moving currents of water, the nutritive value would soon be dissipated. Since the growth of phytoplankton is limited by the availability of nutrients, a method of supplying nutrients to the surface waters is desirable. The development of phytoplankton is limited to the depth of light penetration, as sunlight is necessary for photosynthesis in green plants. Thus there is an ample supply of nutrients in deeper waters which is not being used by the plants. Methods of bringing these nutrients to the surface are being investigated by the Lamont-Doherty laboratories of Columbia University at their Saint Croix station in the Virgin Islands.

Another method of increasing the fertility of the waters would be to control the dumping of waste products into the sea. Indiscriminate dumping leads to overfertilization, resulting in the pollution of water bodies, but the selective dumping of certain elements and organic compounds which could be fully utilized by the natural populations is desirable. The use of waste products to increase production has much to offer, not only in the increased productivity, but also as a method of eliminating some of the pollutants which have been contaminating the waters of the world (Fig. 13-5).

Before 1950, commercial fishing was largely limited to the Northern Hemisphere. Many fishermen tended to stay close to home, although

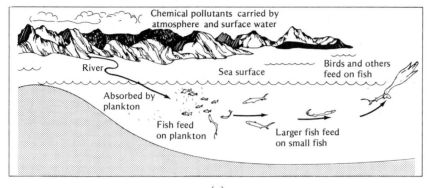

Chemical pollutants carried by
atmosphere and surface water

River

Sea surface

Birds and others
feed on fish

Absorbed by
plankton

Fish feed
on plankton

Larger fish feed
on small fish

(a)

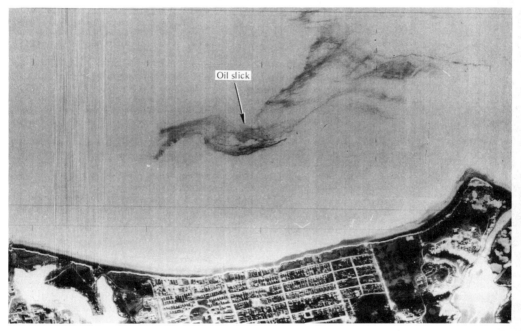

Oil slick

(b)

13-5 (a) The effects of some pollutants in the sea.
(b) An infrared photograph of an oil slick. *(Texas Instruments, Inc.)*

the French, Portuguese, British, and Japanese travel long distances to
fish. Most recently, the Soviet Union has joined the ranks of the
fishing powers and now ranks third in the world. While in 1950 the
southern oceans supplied less than 5 percent of the total world
catch, this same area is now responsible for over 35 percent.

Many countries which lack land for farming depend on their fish yield to provide the necessary protein. Although this is not true of the United States, a reduction in the fish yield would place an additional burden on the agricultural surplus and necessitate a substantial increase in dairy production to replace the protein now fed to livestock in the form of fish meal.

Presently, the oceans yield approximately 55 million metric tons of fish per year, of which 90 percent is finfish (as opposed to shellfish). The remaining 10 percent is composed of whales, crustaceans, and mollusks as well as a variety of marine invertebrates. The annual fish harvest appears to be the only foodstuff that shows a global production in excess of the increased growth rate of the world's population. While these statistics are fact, the trend is for smaller and smaller quantities to be used directly for human consumption. Except for the luxury foods such as shellfish, more and more fish are being converted into fish meal to be fed directly to livestock.

PROBLEMS OF OVERFISHING

It may appear that the world has an overabundant supply of fish, but this impression is misleading. Despite the present and increasing yield of fish, the sea harvest supplies only one-tenth of the world's consumption of protein. The most critical problem facing the world's fish population is that of overfishing. Increased fishing tends to reduce the level of the stock progressively, but a maximum sustainable level may be reached if precautions are taken. It is best to fish the older representatives of a species, leaving the younger ones to replenish the losses. Generally, the loss due to fishing and natural death may be offset by emigration of new forms and the appearance of new offspring. There is, of course, a limit to the amount of fish we can take each year from the natural stock. The extent to which we shall be able to increase our yield will depend on the quantity of unfished stock as well as our efforts of conservation to prevent exceeding the maximum sustainable yield. It appears that if we continue with our present fishing methods, all of the world's fisheries will have been overexploited to some degree.

MANAGEMENT OF FISH RESOURCES

A number of agencies are currently working on the problem of proper management of our fish resources. Working with the statistics of stock potential, they are able to estimate with a fair degree of reliability the growth and mortality rates of fishes. By using these data they are able to guess at the feasibility of fishing a particular stock without

fear of diminishing the reserve. Scientists investigating fish populations project a sustainable maximum yield of between 100 and 200
million tons by 1985. Although this represents a substantial increase
in the available food for future generations, the question still remains
whether it is economically advisable to pursue such a program.

A number of researchers believe that the present course of man's
fish exploitation will soon lead to disaster unless we are more pru-
dent, that is, exercise population control and a more careful harvest
of our present major sources of food. The view is widely held that the
catch from the sea is not inexhaustible and that the oceans have little
potential for increasing the protein yield.

Whether one accepts this view or not, the signs of danger are
unmistakable. For example, the world fish catch has more than doubled since 1950, following an international conference which identified new fishing grounds for exploitation. But the increase can be
shown to be due to an increased utilization of basic fish stocks. For
example, 105 million metric tons of the Peruvian anchoveta were
caught annually in the late 1960s. This represents nearly 20 percent of
the total marine catch in the world. In the late 1950s, only slightly
more than 1 million metric tons of this fish were caught annually.
Thus, we are simply exploiting the same food stocks on an increasing
basis. Obviously, this situation cannot continue indefinitely. The
major fishing grounds exist in only a few parts of the world ocean; the
rest of the ocean is considered by many researchers to be a biological
desert. Further, the major portion of the world's marine catch consists
of only ten species, six species of fish and four species of shellfish.

One proposal for reducing the costs of production and increasing
the yield of available animal protein is to begin using organisms
earlier in the food chain. This would take into consideration the 10
percent efficiency level that exists between successive food-chain
steps. If we catch a ton of fish, we might as well catch 10 tons of the
organisms on which these fish feed. By harvesting the smaller organisms we can reduce the amount of larger fish caught and still achieve
the same purpose. The question remains whether it is more advantageous to turn the collected organisms into fish meal to be fed to
livestock or to allow the food cycle to take its full course and harvest
the marine fish instead.

Thus, we must approach the problem of a rapidly increasing population causing a greater demand for food from two directions. One
approach entails a more efficient and sophisticated means of hunting
and harvesting fish from the already existing supplies.

Successful fishing fleets now rely on such techniques as sonar to
track large schools of fish, techniques particularly successful in de-
tecting concentrated populations of fish at middle depths in the sea.

Sonar allows the vessel to close more quickly on the fish and to determine the precise location of the greater mass of fish.

Satellites and aircraft reconaissance may also be used to track fish and detect new areas of upwelling and possible new sources of food fish in nutrient-rich waters. Methods of detection and an increased knowledge of oceanic circulation will allow improved application of conventional methods of exploiting present stocks of fish caught in relatively small numbers and, concurrently, the detection of new fishing grounds of potentially rich concentrations of fish. At present, the southern waters in the Atlantic and Pacific are largely unharvested except in the regions off the west coasts of South America and Africa. Also, the Indian Ocean has been almost virtually ignored as a source of food.

FISH-FOOD PROCESSING
Another conventional method of increasing the delivery of food to the world is fish-food processing. Curing and freezing fish food allows delivery of this highly perishable food to population masses that ordinarily would not eat fish. For example, the canned-fish industry has increased by a factor of at least 4 in the last generation, and the use of frozen fish has increased by nearly a factor of 8 during the same time. Perhaps eventually changes in eating habits will cause more people to accept seafood as a major part of their diet. The use of so-called *trash fish* as fish-meal products for our table may replace its current use as fertilizer and cattle fodder. And, of course, improvements in netting and trawling may add even further to our source of food.

The second approach entails improvements and increased sophistication in fish farming (Fig. 13-6). This technique may even lead to the utilization of land that is useless for any other purpose at present. Fish farming also yields techniques of breeding, raising, and stocking current freshwater fish supplies, thus raising our potential catch from this very important part of our food supply.

MAINTAINING OUR NATURAL RESOURCES
One additional method of ensuring an increase in future yields is to concentrate on the improvement of our natural fish resources, partic-

13-6   An experimental fish farm in the Philippines. (*FAO-UN*)

ularly the cultivation of highly prized species. The indiscriminate pollution of the sea is posing a threat to the development of productive fishing grounds. The dumping of noxious substances into the sea needs to be regulated and controlled on a worldwide basis or the future of productive fisheries—particularly in the coastal areas—appears grim.

SUMMARY QUESTIONS

1. Although many researchers regard plankton as a major new food source, why does it seem impractical to attempt to remove plankton directly from the sea?

2. What proportion of food actually is turned into body protein at each level in the food chain? How does this loss affect the quantity of plankton that actually translates into table food?

3. What methods are utilized in modern fish farms, or aquaculture?

4. How has indiscriminate dumping affected some potential food sources in the sea?

5. If the world supply of marine food is 55 million metric tons, why is only a small portion used directly as food?

6. Describe three new approaches currently under study or experimentation that might increase our seafood supply.

7. How do we account for the large seasonal variations in primary productivity in some regions of the sea?

8. What are some of the drawbacks prohibiting man from the large-scale development of available waters into aquaculture farms?

9. Discuss the legal and political ramifications of utilizing public-domain areas for fish farming.

10. Describe the various means of increasing the natural fertility of the sea to enhance the production of phytoplankton.

11. What precautions are necessary to prevent overexploitation of the sea's natural fish harvest?

12. Describe several experiments in fish farming and how they improve on the catch from natural sources.

13. Discuss the natural food chain in the sea.

14. Describe an experiment in artificial upwelling.

15. What is the relationship between food intake and increased body weight at each level in the food chain?

16. Which ocean has been almost virtually ignored as a food source?

17. What becomes of most trash fish?

18. How is energy "lost" in the natural food chain?

19. Why are shellfish particularly responsive to aquaculture?

20. Compare the use of sonar to catch schools of fish to the use of helicopters to catch land-dwelling wildlife. The threat of extinction for certain land animals can be extended to pertain to sea animals. Is this true? What can be done about it?

14 | THE SEA AND ENERGY

Our sun is the source of all energy and energy interactions on Earth. As the energy impinges upon the Earth's surface, a portion of it is absorbed and stored by the solid Earth, its water surfaces, and the atmosphere, while the remaining energy is utilized by reactions within the atmosphere or reflected back into space. Of the absorbed and retained portion of the solar energy reaching the Earth's surface, much of it is distributed throughout the oceans and atmosphere by the circulatory patterns set up by the different density currents created by the energy absorption.

The sea and atmosphere are in constant motion. Waves, tides, and convective processes carry energy to all levels of the ocean and throughout all latitudes. All these motions of the sea are the direct result of energy that reaches the surface of the Earth and is absorbed by the land and water. While some of the energy is transferred by reflection from the surface, the absorption is important primarily because it makes energy available to the Earth and its atmosphere through conversion into other forms.

The motion of sea waves leads to the direct emission of radiational energy. The sea receives energy chiefly as shortwave energy such as light waves. The Earth's surface then reradiates the energy into the atmosphere primarily as longwave energy such as heat. About 40 percent of the incoming sunlight is reflected back to space by the atmosphere and never reaches the reservoir of stored energy on the Earth's surface. About 15 percent of the incoming energy, or *insolation,* is absorbed by the atmosphere. Of the total insolation, about

An artist's conception of future oil field exploitation on the continental shelf. *(Reynolds Metals Company)*

40 percent finally reaches the sea surface as direct rays of the sun and as scattered rays from the atmosphere and clouds. Part of the incoming radiation is reflected back into the atmosphere by the sea surface before it is finally absorbed by the water. This reflection varies widely from one place to another; water reflects about 15 percent of the incoming radiation while ice-covered seas reflect from 50 to 80 percent. Thus, about 25 percent of the incoming radiation from the sun actually enters the ocean, where it becomes available for interactions, both chemical and physical, that cause oceanic circulation and variations in water density (see Fig. 14-1).

THE SEA AS A DYNAMIC MACHINE

The high specific heat of water permits a great deal of energy to be absorbed and stored as latent heat in seawater. The extremely moderate rate of change in the temperature of adjacent water masses means that water masses of widely different temperatures can exist side by side in small areas of the ocean. Furthermore, the very low specific heat of the landmasses between the oceans creates marked temperature differences between the land and sea even during the same season in the same latitude of the world.

Air temperatures differ markedly from one place to another, depending upon whether the air mass touches land or water. The temperature of a specific air mass is directly controlled by the emission of energy from the Earth's surface, and the reradiation of energy from land and water varies with the specific heat and ability to absorb energy. Thus, air temperatures throughout the world are found to be quite different when offshore air temperatures and coastal air temperatures are compared.

The latent heat stored by land and water masses is later released directly into the atmosphere as usable energy which changes the character of the air mass. This energy, called *sensible heat,* is then distributed by the air masses which sweep around the globe in various bands of latitude. The major role in heat transport is carried out by the atmosphere, but the oceans play an important role in energy balance by their own circulation.

The amount of energy available to the atmosphere from land and water varies with the ability of these surfaces to absorb and lose energy. Land surfaces, for the most part, are opaque to heat energy. Heat is carried downward through landmasses to a very limited depth, and energy is lost readily and rapidly into the atmosphere. However, water is relatively transparent to energy to great depths.

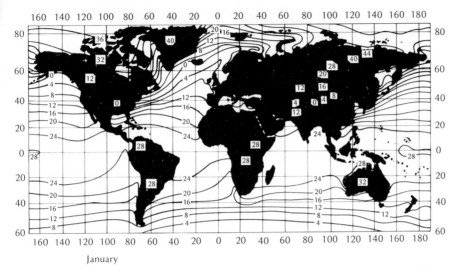

January

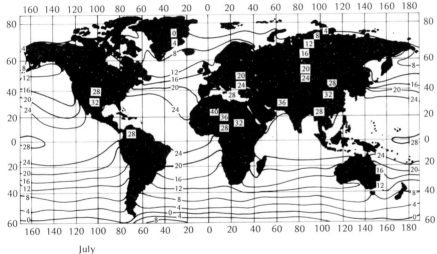

July

14-1 Sea surface temperatures.

In many areas, energy can be found passing through the water to a depth of 100 meters or more. The initial absorption is most important for the purposes of our discussion, but convective mixing in the sea can carry energy downward to great depths in the ocean. Thus, greater amounts of energy are available for later release by the oceans and are stored in a larger mass of water than on land.

AIR MASSES OF THE WORLD

Air masses arise when a specific set of physical characteristics is attained by the air in contact with the Earth's surface. The energy imparted to the atmosphere creates a broad band of circulation bound by relatively large areas of latitude. In each hemisphere, three broad bands of circulation carry these large air masses through large regions of the world both vertically and horizontally. An examination of the general circulation of the atmosphere reveals the major types of circulation created in the atmosphere by the energy exchanges with the Earth's surface. Further, the movements of these air masses are related to the Coriolis effect and pressure systems created by the divergence and convergence of air masses.

However, our weather and climatic patterns are most directly affected by the horizontal movement of secondary air masses beneath the larger patterns of circulation. Initially, the air masses are stable, and little mixing occurs, but they do not remain stable for long, however.

As the air masses form over land or water, they soon begin to travel horizontally under the influence of pressure systems. As they move from one place to another, they carry with them the characteristics imparted to them at their source. As they move, the air masses in turn create changes along the Earth's surface as differing air masses meet. Frontal systems form as different types of air masses join along a frontal line known as a *discontinuity*. Since it is the source region which characterizes air masses, they are named according to the region which gave them birth.

Several classifications are used to place air masses in easily defined groups. Basically, there are four major types of air masses. Arctic (A) and polar (P) air masses are relatively cold and dry. The distinction between the two air masses is small, but the temperature difference between them is measurable. Tropical (T) and equatorial (E) air masses generally are warm and moist; again the temperature differences between the two are small but measurable.

The four basic air masses are subdivided into smaller groups, which are more indicative of their nature and the initial source region. The subdivisions include continental (c), which are dry masses of air originating over land, and maritime (m), which are moist air masses originating over the oceans.

Frequently these air masses also carry other designations denoting more specific physical characteristics. For example, they may be classified as warm (w) or cold (k) when compared to the Earth's surface and Pacific (P), Atlantic (A), and Gulf (G).

These various air masses originate in major source regions of the world. For example, in the Tropics of Cancer and Capricorn we find continental tropical (cT) air masses, which are hot and dry, often associated with the mid-latitude deserts of the world. The maritime tropical (mT) air masses originate over the mid-latitude sea.

The doldrums near the equator produce the maritime equatorial (mE) air masses. Here, there is little land, and major air masses are usually hot and moist.

The extreme North and South Atlantic and Pacific Oceans produce the world's maritime polar (mP) air masses. The continental polar (cP) air masses originate in North America and Eurasia, primarily from source regions between 55 and 65° north latitude. In the southern latitudes, the air masses are primarily maritime polar due to the lack of major landmasses there.

Air Masses and the United States

In the United States, weather is chiefly the result of four major types of air masses (Fig. 14-2). The polar continental air masses from northwest Canada and Alaska near the Arctic Ocean produce cold, clear weather with relatively low humidity. These air masses may

---

14-2 Seasonal air masses of the United States.

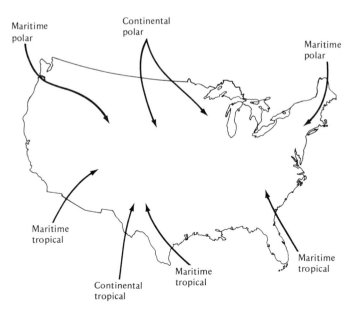

Maritime polar

Continental polar

Maritime polar

Maritime tropical

Continental tropical

Maritime tropical

Maritime tropical

gain moisture as they cross the Great Lakes, however. Radiational loss of heat from the Earth's surface cools the air and produces an air mass which shows a rise in temperature with height. As this situation is exactly the opposite of what would normally occur, temperature inversions (warmer air atop cold air) frequently produce a highly stable air mass. Polar maritime air masses develop in the North Pacific Ocean off the western United States and near Newfoundland, Labrador, and Greenland in the east. These moist air masses gain heat as they encounter landmasses. Their high humidity frequently produces large amounts of rainfall as they are forced to climb mountain ranges such as the Cascades and Sierra Nevada on the West Coast. One result of the tropical continental air masses arising in Mexico and sweeping into the hot and dry Southwest is the great American desert. Near the same region of our country, tropical maritime air masses move into the Gulf Coast states and the Southeast from the Gulf of Mexico, the Caribbean, and the South Atlantic Ocean. On the West Coast, fogs result as warm, moist air moves out of the South Pacific across the cold landmass of California.

THE HEAT BUDGET OF THE OCEANS

Although the annual quantity of solar energy received by the Earth remains amazingly constant over long periods of time, the energy received in different places or climatic regimes varies widely from one season to another. The energy absorbed at the Earth's surface is not constant throughout the world because of the circumstances responsible for seasonal changes.

If the planet on which we live rotated with its axis exactly perpendicular to the surface of the sun, we would experience no seasonal changes during the course of a year. However, the Earth's axis is tilted slightly, so that our rotation occurs around a polar axis that is about 23.5° from the vertical. Further, the tilt of the axis is oriented toward the same point in the heavens, approximately toward Polaris, the pole star. As a result, the orientation of the Earth's surface and axis constantly shifts slightly in respect to the sun.

In the Northern Hemisphere June 21 is the longest day of the year. At this time, the Earth's northern regions are pointed toward the sun, the Northern Hemisphere spends the greatest amount of time in sunlight, and the North Pole experiences its 6-month summer (Fig. 14-3). December 21 is the shortest day, the point in our orbit when the northern regions are tilted away from the sun. These two dates represent the summer and winter solstices, respectively.

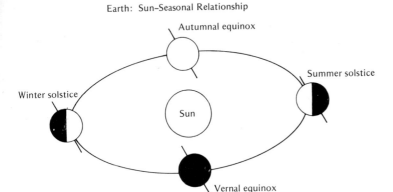

Earth: Sun–Seasonal Relationship

14-3   Seasonal relationships of the Earth and sun.

On March 21 and September 22 the days are approximately as long as the nights. These two dates represent the spring (vernal) and autumnal equinoxes (named from Latin words meaning "equal night"). During these periods the Northern and Southern Hemispheres are equally illuminated.

Because of the constantly shifting orientation of the Earth in respect to the sun, solar energy strikes the Earth's surface at different angles throughout the year. On June 21, the sun's rays are perpendicular to the Earth's surface in the Northern Hemisphere at the Tropic of Cancer, 23.5° north latitude. On December 21, the rays of the sun are perpendicular to the Tropic of Capricorn, 23.5° south latitude. Thus, the rays of energy enter the Earth's atmosphere at a more direct angle during the summer and more obliquely during the winter.

The variation in the angle of incoming solar radiation alters the concentration of energy reaching the Earth's surface. In the winter months, the energy entering at an oblique angle is spread over a greater area as it passes through a greater portion of the atmosphere. The same quantity of energy in the summer passes directly through a smaller portion of the atmosphere, is spread less, and therefore is more concentrated. This change in the angle of incoming radiation produces a great variation in the amount of energy absorbed by the Earth's atmosphere, land, and water.

Thus, the amount of energy received by the Earth's surface depends upon the latitude at which the energy enters the atmosphere, the seasonal effects on the energy received by a particular locale, and the state of the atmosphere, that is, cloudy or clear.

The variations between latitudes and seasons are rather marked in respect to the amount of energy received by the Earth's surface. The middle latitudes of the Earth receive approximately 600 calories per square centimeter of surface per day during the summer months. In the winter, these same latitudes receive as little as 100 calories.

Further, the balance of energy varies with the season. In the Northern Hemisphere, the gain in energy exceeds the loss during the months from March to September. From September to March, the loss generally exceeds the gain in energy. The ocean slowly warms during the first period and slowly cools during the second period in the Northern Hemisphere.

Within 30 to 40° latitude on either side of the equator, the annual energy gain exceeds the loss; and within 30 to 40° of either pole, the annual loss of energy exceeds the gain.

Thus, the balance of temperature achieved throughout the world is accomplished through atmospheric and oceanic circulation. Each medium achieves a relatively constant yearly temperature in all latitudes by carrying energy from regimes of excess energy toward those lacking in energy during each season.

EVAPORATION AND RADIATIONAL EFFECTS

The energy absorbed by the ocean surface reenters the atmosphere by one of two processes: (1) latent heat stored in water as it evaporates into the air and (2) direct radiation into the atmosphere. Each process, directly or indirectly, allows for later utilization of energy as sensible heat. This sensible heat is responsible for changes in the atmosphere and the motion of air masses.

EVAPORATION

Evaporation of water occurs throughout the world from both land and water surfaces. The oceans lose water to the atmosphere; it is a rapid loss, allowing energy to be carried into the atmosphere at a rate of 600 calories per gram of water. Later, when condensation of the water vapor occurs, the energy is lost to the molecules of air as sensible heat. The average energy gain is about 180 calories of heat per square centimeter of atmosphere.

Evaporation and loss of heat do not occur as rapidly from land as from the sea. On the continents, evaporation occurs most readily where soil moisture is available, and this varies widely from one season to the next in all parts of the world. Land temperatures change more swiftly than temperatures in the oceans.

Since evaporation over the oceans occurs constantly, energy is always available to the atmosphere in large quantities from the ocean. In the winter months, oceanic surfaces warm the air slightly and add moisture to the air.

In general, a higher rate of evaporation of moisture occurs on the eastern portions of mid-latitude continental regions. The rate of evaporation is enhanced by the cold, dry winter winds moving from the continents onto the water surfaces.

The ocean actually loses more moisture through evaporation processes than it gains from precipitation. The water deficit is made up by runoff of surface water from rivers emptying into the oceans.

Although world temperatures decrease toward the polar regions, the changes are highly irregular because of the effect of ocean currents and the interference of landmasses. The change in temperature characteristics is greater over landmasses than over water. Precipitation differences are similar because the rate of evaporation changes. Precipitation decreases toward the interior of a continent as a result of continental winds in the interior.

RADIATION

Large quantities of energy are lost from the oceans by radiation back into the atmosphere. This energy warms the air and causes the creation of low-pressure zones, or cyclones. In extreme situations, cyclones may build into immense storm systems, which depend upon oceanic heat for their development and persistence. In extreme situations, hurricanes, which are characteristic of tropical seas, result (see Fig. 14-4). These storms are called *hurricanes* in the Atlantic, *typhoons* in the Pacific, and *cyclones* in the Indian Ocean. The major hurricane spawning centers throughout the world are in the South Atlantic Ocean, in the Caribbean Sea, west of Mexico and Central America, the South China Sea, the Bay of Bengal, South Indian Ocean, and near the Philippine Islands (Fig. 14-5).

Cyclones are generated in the intertropical convergence zone. In these regions the ocean supplies large amounts of heat into the atmosphere, and in the Northern Hemisphere, the winds created move upward and surface winds begin to spiral inward toward the eye of the storm (see Fig. 14-6).

As upper-level air sinks toward the surface, it gains heat from the water below and warms rapidly, creating a strong convection cell of circulation. Rain soon begins to condense and fall, releasing more latent heat, which becomes available to the system. The wind flow then intensifies, and the winds may eventually exceed 100 miles per hour in a well-developed storm system.

14-4 Hurricane Gladys, a 1968 satellite photograph. *(USNOO)*

As water continues to evaporate from the surface, warm, moist air is trapped in the lower levels of the storm system. As convection continues, the low pressure in the eye builds. The winds continue to spiral around the eye of the storm, and the convection intensifies.

In the Northern Hemisphere, the storm system begins to travel in approximately a northwest direction. The speed is about 10 to 30 miles per hour. However, as the storm crosses landmasses, the hurricane enlarges in scope, weakening the system. It needs a constant supply of energy from the warm seas to keep going, and the contact with land begins the destruction of the storm. The loss of energy input and friction with the land surface slow the storm, and unless it crosses onto a warm ocean surface again, it eventually dissipates.

THE DISTRIBUTION OF ENERGY IN THE SEA

The heat balance in the oceans is affected by a number of agents, which tend to add heat to the ocean water in some situations and remove it in others. The heating processes involved in the ocean are varied and include the following:

Ocean Wave Forecasts/National Meteorological Center

Weather and sea state observations are made by ships and ocean stations. ➤

Data are transmitted direct to computers at NMC. ➤

Computer output is fed into automatic chart drawer.

Newly drawn chart is run through facsimile machine for transmission to coastal and high-seas forecast centers. ➤

Marine advisories are sent by teletype to marine radio stations for broadcast. ➤

Ships at sea receive forecasts by radio and use them to navigate more safely.

14-5 Ocean-wave forecasting sequence. *(NOAA)*

14-6 The hurricane source regions of the world.

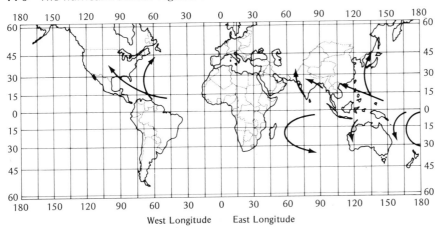

West Longitude   East Longitude

Radiation energy enters the water from solar energy directly impinging on the Earth's surface and reflected from the atmospheric cloud cover.

Water vapor and the formation of dew release heat energy on the ocean surface. This energy is absorbed and enters the heat cycle of the sea.

The atmosphere also conducts heat, and the contact with the sea surface allows for some of this energy to be absorbed by the water body.

Conduction also plays a role in the release of heat from the rocks in the ocean basins. Although this factor is minimal, some heat probably enters the bottom waters from magmatic fluids beneath the ocean floor. Heat from radioactive processes plays a major role in the formation of magma beneath the Earth's surface.

Mechanical energy from wind motion and current movement is converted into heat energy in the waters of the ocean. The movement of air and water over and within the oceans creates mechanical energy, which is frequently converted into sensible heat.

Finally, the activity of living organisms and chemical reactions in the water produces heat which is absorbed by the ocean waters.

Ocean water is cooled by several major processes that produce a loss of heat to a variety of media. Among these processes are the following:

Radiational loss of energy from the surface of the sea to space and the atmosphere causes a significant loss of heat throughout the world's oceans.

A large quantity of heat is lost as latent heat through evaporation of water from the surface of all the water bodies of the world.

In regions where ocean temperatures are high, some energy is lost into the atmosphere through simple conduction of the heat upward through the atmosphere-ocean interface.

A small portion of energy is undoubtedly lost back to the ocean-floor rocks by conduction, just as heat may be gained in the same fashion.

Thus, the oceans, although their role in heat distribution is less than that of the atmosphere, play an important part in the absorption, loss, and balance of heat throughout the world. The sea is in a state of constant flux, and circulation and physical processes of change in the sea create a situation which, in conjunction with the atmosphere, allow for the equitable temperatures that we find in equilibrium throughout the world and over the millennia.

SUMMARY QUESTIONS

1. How do the atmosphere and ocean respectively affect insolation?

2. How are air masses affected by water masses? What relationship between these two portions of our planet has a direct bearing on weather and climate?

3. List and explain the four basic air masses and the subdivisions of each.

4. What are the major air masses of the United States, and during which season does each prevail? Include the major physical characteristics of each.

5. Describe the Earth-sun relationship during each of the four seasonal positions of the Earth's revolution around the sun.

6. How does evaporation affect the heat contained within the atmosphere and the ocean?

7. Where are the world's major hurricane source regions? Why are these places the center of hurricane activity?

8. List and describe the various factors that affect the heat budget of the ocean.

9. How are seasons affected by the Earth's relationship to the sun? What actually causes seasons?

10. How do different surfaces change the reflection and the receipt of energy by the Earth's surface?

11. How do water and land affect the absorption of heat with depth?

12. Describe a temperature inversion. What does it do to the weather situation in a particular place?

13. What factors work together to control the amount of energy received by the Earth's surface?

14. Explain the difference between latent heat and sensible heat.

15. How does evaporation affect the heat contained by a substance? How does this affect sea energy?

16. How does energy from the sea produce and feed a storm system such as a hurricane?

15

THE OCEANS AND CLIMATE

Our daily weather changes and long-term climatic averages are the direct result of water and land relationships throughout the globe. While many things are responsible for the weather and its cumulative climatic regimes, the four combinations of factors are mainly responsible. *Latitude* position is a major reason for differences in weather and climate since it is related to the Earth-sun orientation in space and the amount of energy available to the surface of the Earth at different times during the year. *Oceanic circulation* and the nature of ocean currents are chiefly responsible for climatic regimes along the coastlines of continents and, of course, across island chains in oceans. The presence of *landmasses* around the globe has an important effect on the circulation of air masses and winds; in addition, the *topography* of the landmasses across which the air moves plays an important role in modifying the air masses as they sweep around the globe. These various effects have altered with time, and their influence on our weather and climate has changed during the evolution of our planet. As a result, the weather and climate of our planet have had both minor and major shifts throughout time.

OCEAN CURRENTS AND CLIMATIC INFLUENCES

The heat distribution in the ocean as it is affected by various ocean circulatory patterns is of great importance; in some cases, man's survival may depend upon the particular set of circumstances created by the circulating waters. For example, the Gulf Stream begins

U.S.S. *Missouri* in a fogbank. *(NOAA)*

its journey near Florida with waters that average about 23°C (80°F). As the waters move north and then swing northeast toward Europe, the accumulated energy is lost into the cooler water encountered by the original Gulf Stream (Fig. 15-1). In addition, the atmosphere also receives energy from the North Atlantic Drift, as this tongue of water is now called. This energy exchange in northwest Europe creates a warmer climate than would be expected in the region of Denmark and the Scandinavian peninsula. Since this water prevents the surface of the extreme North Atlantic Ocean from freezing over, the fishing industry of Scandinavia is able to rely on the large fish population in the North Atlantic Ocean.

The pattern of water flow was interfered with during the first part of our present millennium, and the waters of the North Atlantic did, in fact, freeze over in the region of Greenland and Iceland. This played a role in reducing the influence of the early Vikings in the New World.

In most areas of the world, fertile fishing grounds coincide with regions where upwelling produces a current of nutrient-rich cool water. Upwelling is the direct result of a meteorological phenomenon that affects the movement of coastal waters on the western side of continents in the temperate zones of the world, the coasts of California, southwestern Africa, western Australia, and western South America being the most notable.

15-1   Temperatures associated with the Gulf Stream in the Atlantic Ocean.

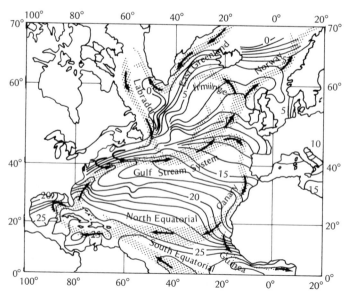

These regions of upwelling result from coastal winds moving parallel to the shore and causing the warm surface water to move offshore. Thus, cold, dense, bottom water moves up to replace the water moving out to sea. The cold water holds larger quantities of nutrients than the warm water it replaces, and fish flourish in it. In South America the process frequently slows in middle to late winter with the slackening power of the southern winds. When the process slows, the fish die because the cold bottom water no longer reaches the surface and it has been replaced once more by a warm surface flow.

Regions that experience cold, coastal waters also are noted for foggy coasts. Here the prevailing warm, maritime winds cool as they cross the cooler coastal water. The condensing water vapor from the wind creates the fog; the current produces a cooler climate than might be expected in the coastal region immediately adjacent to the current.

Cool water then also serves as a means of reabsorption of atmospheric energy. In the Gulf Stream, the warm current eventually swings south along the coast of Europe. Now called the Canaries Current, it cools the area around it as it absorbs energy in its journey toward the equator.

Frequently there is a great disparity between the climatic regimes along opposite coastlines and the interior of the same continent. The differences between coastal temperatures and the contrast between them and interior regions are the direct result of differing currents and the type of winds produced in the interior and along the coastlines.

MARINE AND CONTINENTAL CLIMATES

The major wind systems of the world are modified as they circle the globe and blow over land or water (Fig. 15-2). For example, the powerful Trade Winds move across the globe from the high latitudes, beginning about 30° north and south, and sweep across toward the equatorial doldrum zone. The Trade Winds begin as cool, dry winds but soon change as they cross the oceans just north and south of the equator. The moisture they absorb is lost in the tropics during the tropical rainstorms these winds bring to regions near the equator.

North of the Trade Winds, the surface winds blow from the west to the east in two broad bands that parallel one another in the 30 to 60° band of latitude in each hemisphere. The Westerlies, as they are called, drive weather patterns across the temperate zones from west to east, just the opposite of the weather movement found in the Trade Winds.

The west coasts of continents in the Westerlies are affected prima-

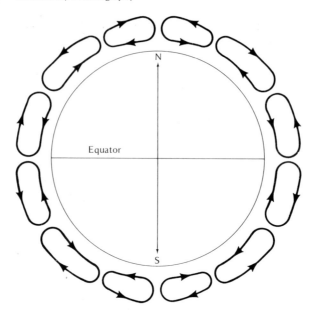

15-2   Worldwide surface-wind circulation.

rily by maritime winds from the ocean basins. The eastern sides of continents in the Westerlies are affected primarily by continental winds, and therefore are much more susceptible to drought than the west coasts.

The movement of tropical storms is also controlled by oceanic circulation and wind patterns. As cyclonic storms grow and build in the tropics, they begin to move north and enter the region of the Westerlies. The storms generally follow the movement of warm ocean currents into these high latitudes, but once they enter the high latitudes, the storms cross land or cold water and their strength is dispersed.

The prevailing winds in each hemisphere are modified by the respective ocean currents, and as a result the climatic conditions on each coastline differ. North of about 40° latitude the east coasts are relatively cold compared to the milder winters found on the western sides of the continents. South of 35° latitude the opposite situation prevails in the summer months. The east coasts are warmer in summer than the west coasts in the same latitude.

Several regions that are noted for fogs throughout much of the year also are affected by ocean-current modification of winds. The production of fog across the region affected by the Peru Current has already been discussed. Fog produced in this manner is called *advection fog*.

Continental climates generally are dry, due to the continental winds from the interior of landmasses that produce the climate. These climatic regimes have very few cloudy days and only light rainfall during most of the year. Arid and semiarid regions, like those found in deserts and steppes, are particularly dry due to the presence of dry continental winds throughout the year.

Unlike the mid-latitude regions tropical continents experience a small range in temperature when winter and summer temperatures are contrasted. The polar regions experience long and severe winters with extremely short, cold summer seasons. The climates along coastlines are less severe when different seasons are compared, thanks to the moderating influence of ocean currents. The seasonal temperature ranges are less than in the interior, a feature that is particularly noticeable on islands in the ocean, compared with larger continental landmasses.

Along coastlines the effect of the adjacent body of water creates a seasonal time lag between the expected onset of winter and its actual occurrence. Littoral coastal climates are the result of currents and their effects on winds. These climatic regimes are midway between a true marine and a continental climate.

LATITUDES, CURRENTS, AND CLIMATES

The climates found throughout the world are caused chiefly by the prevailing winds blowing across vast areas of the globe. The winds carry weather systems along beneath them and distribute energy throughout the atmosphere as they interact with the ocean surface (Fig. 15-3).

In general, warm ocean water moves poleward along the western borders of the oceans. The colder polar water moves toward the equator along the eastern sides of the ocean basins, particularly in the subtropics.

In the high latitudes, warm water flows toward the poles along the eastern sides of the ocean basins. Cold polar water moves toward the equator along the western sides of the basins.

The United States experiences more moderate winter and summer temperatures along the coastlines than in the interior. The range of temperatures is greater inland in the same latitude than on the coasts. For example, San Francisco experiences temperatures which range in the mid sixties at the same time that the Sacramento Valley may

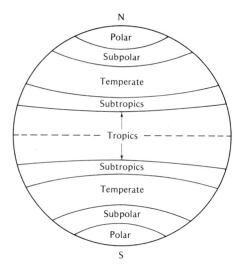

15-3   General climatic regions of the world.

reach temperatures in excess of 100°F. The coast city of San Francisco is cooled by maritime winds, but the more eastern valley is scorched by dry, continental winds.

In addition, temperatures over the open ocean differ greatly compared with continental temperatures. Temperature changes are greatest in a south-to-north direction over the continents in midwinter in January. The temperature changes over water during this same season are least over the open sea. In July, temperature regimes show that the ocean has warmer zones than the landmasses.

Thus, the various latitude zones of the world can be classified according to a few major climatic regimes. The basis for the classifications essentially depend upon the temperature regimes and source regions for the prevailing winds in each zone.

The tropics are those regions bordered by areas of the globe averaging 64°F in the coolest months of the year. In this region, the intertropical convergence zone of winds creates the equatorial low pressure called the doldrums. In the tropics, the temperature range throughout the year is small, and the prevailing winds are the primary reason for the mixing of air masses. The entire region is often characterized by heavy rains and thunderstorm activity.

The middle latitudes generally have an average temperature in excess of 50°F during their warmest months. This region is the area where cold and warm air masses meet and interact with each other. Most of the world's major deserts are found in the mid-latitudes due to the continental winds and topographic effects on the winds. Air masses mix as the result of convective circulation of air along the line of a frontal system. There is a general gain of heat energy for most of the year in the mid-latitudes, and the seasonal temperature ranges are greater than those of the tropics. The west coasts of continents have mediterranean climates characterized by wet winters and dry summers. The east coasts are continental in nature and experience greater temperature ranges than the west coasts. The prevalence of Westerlies in this latitude zone means that temperatures and rainfall are highly variable in nature.

The polar regions are marked by scrubby vegetation and little life. Here, little precipitation falls, and the summer months are extremely short with temperatures that seldom rise above 50°F. The polar regions are the source regions for cold winds and currents. Strong winds and convection cause a good deal of mixing of the air, and the temperature ranges throughout the year are greater than those in the tropics. However, the polar temperature ranges are still less than those found in the mid-latitudes.

Thus, in general, the Westerlies modified by the ocean currents of the world create marine climates on the west coasts of continents and continental climates on the east coasts in the mid-latitudes. The Trade Winds create east-coast marine climates and continental west-coast climates in the low latitudes (see Fig. 15-4).

CYCLICAL CHANGES IN CLIMATE

Although the climates of the world are amazingly stable when viewed over the millennia of Earth history, there is evidence of fundamental shifts in the climatic conditions on our planet. Evidence from fossils and remains of human habitation reveal many changes in the Earth's climate, both short-term and long-term. Parts of the Earth have alternated between conditions of great and small amounts of precipitation. However, the changes occur slowly and take a long time. One of the basic changes and the reasons for it still confound scientists investigating paleoclimatic evidence, namely, the onset of the great glacial epochs which have periodically caused large areas of the globe to be covered by immense ice sheets.

GLACIATION AND OCEANIC CIRCULATION

Many theories have been put forth to explain the development of glacial periods on the Earth. One of the most interesting, the Ewing-Donn hypothesis, sees the glacial periods as a consequence of changes in oceanic circulation in the Northern Hemisphere.

(a)

(b)

(c)

15-4   (a) A polar environment. [*Standard Oil Company (N.J.)*]
(b) A temperate zone environment. *(USDA)*
(c) A desert environment. *(USDA)*

At present the Earth is covered with ice sheets that cover about 10 percent of the total surface. This percentage is much less than has occurred in the past during the various Ice Ages. However, during some of the interglacial periods, even less ice was present on the Earth than there is now. The slowly increasing and decreasing levels of ice on the Earth result in decreasing and increasing levels of the sea. Since the ice sheets grow as the result of increasing precipitation at the poles, the ice must increase at the expense of the oceans as they lose water through evaporation. Conversely, as the ice sheets melt, the water must be returned to the sea as increased runoff from rivers on land.

The Ewing-Donn hypothesis presents the possibility that Ice Ages are initiated by changes that affect the circulation of water between the warmer North Atlantic Ocean and the Arctic Ocean. At present, the warm Atlantic water moves into the Arctic and keeps the Arctic relatively ice-free. The theory assumes that the surface of the Arctic Ocean is the chief contributor of evaporated water to the north polar region. The subsequent condensation of the water vapor contributes to the buildup of polar ice.

Over a long period of time the constant evaporation of water from the Arctic Ocean causes its level to drop. The water vapor continues to be locked up as ice at the pole.

A narrow rock sill between Norway and Greenland forms a separation between the Arctic Ocean basin and the North Atlantic Ocean (Fig. 15-5). As the ice sheet increases in size and begins to advance southward, the lowered sea level eventually reaches the upper level of the rock sill between the two oceans, and circulation of Atlantic water into the Arctic becomes slowed and eventually ceases.

Warm water no longer enters the Arctic, and the surface freezes over. As the ice builds, it also causes a further drop in air temperature and the Ice Age reaches its peak.

15-5   A cross section of the Arctic Ocean rock sill near Spitsbergen separating the Arctic and Atlantic Oceans. *(After Ewing and Donn)*

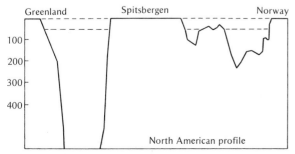

North American profile

However, the glacial advance is also self-destructive. Once the Arctic freezes over, the source of evaporation for the precipitation of snow ceases and the ice can no longer grow and advance. This begins the stage of melting of the ice sheet. The meltwater runs into the sea, and the sea level slowly begins to rise once more.

As the ice melts, the air temperatures begin to rise slowly again. The increasing level of the sea causes the Atlantic flow to move across the rock sill between the two ocean basins, and the Arctic is again freed of its ice cover. The Ice Age comes to an end, but a new cycle starts the succession of events slowly toward a new Ice Age.

There is evidence that such changes do occur on a small scale as well as on a large scale. During the last Ice Age, which occurred over the last 2 million years, the sea level dropped by as much as 300 to 500 feet. The present continental shelves were probably exposed as our ancient coastlines.

In the tenth century, a minor advance of ice in the North Atlantic caused the waters between Greenland and Scandinavia to freeze over. This cut off the Viking colonies from their home territory. The colonists were not able to receive aid and supplies from home, as the Eskimos moved back in and apparently destroyed the colonists who had originally pushed them out of their own lands. The loss of herring fishing, one of the major sources of food for the Vikings, also helped to destroy the Viking culture as a major power in the North Atlantic Ocean.

PAST CLIMATES OF THE EARTH

Much research has been conducted into past climatic changes on Earth. Although we do not have a complete picture, many parts of the puzzle are clear at the present time. They reveal that numerous short- and long-term changes have occurred and that our present climatic regimes have varied widely even during the last millennium.

It is currently believed that about 3000 B.C. worldwide temperatures were about 1.5 °C (3 °F) higher than at present. The seas were about 1 to 2 meters (3 to 6 feet) higher than they are today. The desert regions of the Mideast received more rainfall than at present, and farming was more profitable.

A series of minor Ice Ages have occurred periodically; for example, small glacial advances in 500 B.C. and others in the seventeenth and nineteenth centuries. Since then the seas have risen at a rate of about 5 to 6 centimeters (2 to 2½ inches) per thousand years.

Some repetitive tendencies seem to occur in the climatic shifts, but the overall pattern has not yet been worked out.

Several reasons have been offered for these changes. Most revolve around fluctuations in the relationship of the Earth and sun or changes in the surface character of the Earth.

Some astronomers believe there may be fluctuations in the expansion and contraction of the outer portions of the sun. Such changes would be due to the concentration of heavy elements created in the sun by the nuclear-transformation processes taking place in its interior. This fluctuation would result in changes in the energy output and thus affect the energy received at the Earth's surface.

Others have suggested that changes have taken place in the eccentric ellipse of the Earth's orbit around the sun. If the Earth moved into an orbit closer to or farther from the sun, the distance from the Earth to the sun would change and insolation would be altered.

Evidence reveals that worldwide temperatures have varied through a range from about 42 to 72 °F. At present, average worldwide temperatures are about 58 °F. This would imply a very fundamental change in our basic relationship to the sun or in our atmosphere.

One of the major components of the atmosphere on the present Earth, carbon dioxide gas, is thought to be critical in the absorption of energy by the atmosphere. Changes in the percentage of this gas may be brought about by changes in the number of volcanic eruptions and the amount of animal life present on the Earth's surface. These changes may account for an alteration in the amount of energy absorbed by the atmosphere.

There is some recent evidence to support this last hypothesis. From 1880 to 1940, an increase in worldwide temperatures was noted. This increase amounted to 1 °F (about ½°C). It is thought that this temperature increase is due to the excess carbon dioxide thrown into the atmosphere by industrial processes. However, from 1940 to the present, temperatures have decreased by about ½°F.

At first glance the situation seems contradictory. However, it is believed that the decrease in temperature is the result of the reflection of energy back into space by the long-term accumulation of particulate matter in the atmosphere (smoke, dust, and other unburned particles) (Fig. 15-6). Thus, atmospheric pollution may produce both a heating and a cooling effect on the Earth's surface. In the long run, this effect on solar energy would change the solar energy impinging on the ocean's surface.

Still another idea is that changes in the topography of the land and shifts in the amount of land and sea surface bring about major changes in climatic patterns. We have already seen that land and sea surfaces affect the surface temperatures of the Earth in different ways. It is believed that geologic uplift of new landmasses or the downdropping of landmasses slowly alters the amount of land and

15-6   Photochemical smog in Washington, D.C. *(USDA)*

---

sea surfaces in contact with the atmosphere. These changes affect the amount of absorbed energy and the later reemission of the energy in different latitudes of the globe.

These alterations affect climate over long periods of time. Changes in the position of the drifting continents would also play a role in climatic patterns.

Thus far, researchers have managed to develop an approximate pattern of changes in worldwide climate on a long-term and short-term basis. These changes are shown in Tables 15-1 and 15-2.

Throughout most of the Earth's history, climates have tended toward warm periods. The reconstruction of past climates is complicated by geological uplifts, erosion, and other changes in topography. In addition, the fossil evidence is not complete, but basic fundamental changes in climate do occur.

TABLE 15-1   Past Climates of the Earth

Era	Years in millions	Period	Worldwide climate
Cenozoic		Quaternary:	
	0–1	Pleistocene	Four Ice Ages, temperate-zone glaciers
	1–13	Pliocene	Cool
	13–25	Miocene	Moderate
		Tertiary:	
	25–36	Oligocene	Moderate to warm
	36–58	Eocene	Moderate, then warmer
	58–63	Paleocene	Moderate
Mesozoic	63–135	Cretaceous	Moderate
	135–181	Jurassic	Warm
	181–230	Triassic	Warm
	230–280	Permian	Glaciation, then moderate
	280–345	Carboniferous	Warm, glaciation in Southern Hemisphere
Paleozoic	345–405	Devonian	Moderate, then warm
	405–425	Silurian	Warm
	425–500	Ordovician	Moderate to warm
	500–600	Cambrian	Cold, then warm
		Precambrian	Glaciers in both hemispheres

SOURCE: W. D. Sellers, "Physical Climatology," University of Chicago Press, Chicago, 1965.

TABLE 15-2   Recent Climatic Events

Years	Location	Trend
B.C.:		
9000–6000	Europe	Cooling to 7000; ice increased in Scandinavia
6000–2500	North America and Europe	Cool and dry; then warm and wet; droughts by 3000
2500–500	Northern Hemisphere	Warm and dry; then intermittent heavy rain followed by European drought
500–0	Europe	Cool and moist; glaciers in Scandinavia
A.D.:		
300–600	United States	Southwestern drought; Alaskan glaciation
600–800	United States and Europe	Advance of Alaskan glaciers; Europe dry; Near East drought
800–1000	Europe	Scandinavian glaciers retreated; North Africa cold
1000–1200	United States	Ice in the west; Alaskan glaciers advanced
1200–1500	United States	West wet followed by drought conditions
1500–1900	United States and Europe	Cool and dry; European glacial advance in 1550–1680, in 1740–1770, and in the 1800s; North American glaciers advanced in 1700–1750, following a drought in the southwest during the 1500s
1880–1940	Both hemispheres	Temperature increase of 1°F; glaciers reduced
1920–1958	United States	Southwestern dry spell
1942– present	Both hemispheres	Temperatures decreased by ½°F; glaciers stable

SOURCE: W. D. Sellers, "Physical Climatology," University of Chicago Press, Chicago, 1965.

SUMMARY QUESTIONS

1. How are weather and climate affected by latitude, oceanic circulation, landmasses, topography, and air circulation?

2. Why are there great differences in climate between coastal areas and inland areas at the same latitudes? How does climate differ on east and west coasts in the temperate zone of the Earth? Why?

3. How are advection fogs related to ocean circulation?

4. Why are there seasonal time lags between ocean and land temperatures? Briefly describe the differences.

5. What types of major climatic cycles have we noted during the Earth's history? In what way might they be related to ocean circulation?

6. What types of minor shifts in climate occurred in the last few millennia? Might man be affecting climate today? In what ways?

7. Describe how the entire gyre starting with the Gulf Stream has various effects on the United States and Europe.

8. Where are the major regions of upwelling in the world? What meteorological phenomenon are they associated with?

9. How do the Westerlies and Trade Winds affect the weather patterns in their respective portions of the globe?

10. List the general physical characteristics for the tropical, temperate, and polar zones.

11. Describe the events of the Ewing-Donn hypothesis to explain the onset of Northern Hemisphere glaciation.

12. List at least three other reasons for glaciers that have been proposed. Include one that deals with changes in atmospheric composition.

13. How have temperature trends appeared through most of the Earth's history? What factors complicate reconstruction of past history?

APPENDIX TABLES

TABLE A-1   The Major Ocean Trenches

Name	Ocean	Depth in meters	Length in kilometers
Philippines	Pacific	11,516	1,200
Marianas	Pacific	11,033	2,500
Tonga	Pacific	10,882	1,250
Kuril-Kamchatka	Pacific	10,542	2,150
Kermadec	Pacific	10,047	1,200
Bonin	Pacific	9,810	500
Puerto Rico	Atlantic	9,200	800
New Hebrides	Pacific	9,165	500
Solomon	Pacific	9,103	300
Yap	Pacific	8,527	400
New Britain	Pacific	8,320	450
South Sandwich	Atlantic	8,263	1,200
Peru-Chile	Pacific	8,055	1,900
Palau	Pacific	8,054	200
Diamantina	Indian	8,047	160
Aleutian	Pacific	7,679	3,300
Ryukyu	Pacific	7,507	700
Cayman	Caribbean Sea	7,491	900
Java	Indian	7,450	2,400

SOURCE: Adapted from Andre Cailleux, "Anatomy of the Earth," World University Library, McGraw-Hill Book Company, 1968.

TABLE A-2   Relative Size of the Oceans

Ocean	Surface in square kilometers
Pacific	180,500,000
Atlantic	92,200,000
Indian	75,000,000
Arctic	14,000,000

SOURCE: Adapted from Andre Cailleux, "Anatomy of the Earth," World University Library, McGraw-Hill Book Company, 1968.

TABLE A-3   The Ten Largest Lakes

Name	Area in square kilometers	Location	Depth in meters
Caspian Sea	394,000	Eurasia	980
Superior	82,410	United States	406
Victoria	69,485	Africa	81
Aral Sea	66,460	Asia	68
Huron	59,830	United States	229
Michigan	58,020	United States	281
Tanganyika	32,890	Africa	1,435
Great Bear	31,790	Canada	82
Baikal	31,500	Asia	1,741
Nyasa	30,000	Africa	706

SOURCE: Adapted from Andre Cailleux, "Anatomy of the Earth," World University Library, McGraw-Hill Book Company, 1968.

TABLE A-4   The Largest Islands

Name	Area in square kilometers
Greenland	2,175,000
New Guinea	820,000
Borneo	744,000
Madagascar	594,000
Baffin	476,000
Sumatra	473,000
Honshu	228,000
Great Britain	230,000
Ellesmere	212,000
Victoria	212,000
Celebes	188,000
South Island	152,000
Java	124,000
North Island	115,000
Cuba	114,000
Newfoundland	111,000
Luzon	107,000
Iceland	104,000
Mindanao	95,800
Hokkaido	89,900
Ireland	84,400
Novaya Zemlya	79,000
Sakhalin	75,400
Haiti	74,100
Tasmania	67,960

SOURCE: Adapted from Andre Cailleux, "Anatomy of the Earth," World University Library, McGraw-Hill Book Company, 1968.

TABLE A-5   The Largest Seas

Name	Area in square kilometers	Average depth in meters	Maximum depth in meters
Caribbean	2,750,000	2,490	7,491
Mediterranean	2,500,000	1,485	4,901
Bering	2,270,000	1,436	3,961
Gulf of Mexico	1,540,000	1,512	4,376
Okhotsk	1,530,000	838	3,379
East China	1,250,000	188	2,681
Hudson Bay	1,230,000	128	258
Japan	1,010,000	1,350	3,617
North	580,000	94	725
Black	460,000	1,100	2,246
Red	440,000	491	2,359
Baltic	420,000	56	459

SOURCE: Adapted from Andre Cailleux, "Anatomy of the Earth," World University Library, McGraw-Hill Book Company, 1968.

TABLE A-6   Equivalents

Centimeter	= 0.3937 inch
	= 0.0328 foot
Meter	= 39.37 inches
	= 3.28 feet
Inch	= 2.54 centimeters
	= 25.4 millimeters
	= 0.08 foot
Foot	= 30.48 centimeters
	= 0.3048 meter
	= 0.16667 = $\frac{1}{6}$ fathom
Yard	= 91.44 centimeters
	= 0.9144 meter
Fathom	= 6 feet
	= 1.828 meters
Kilometer	= 0.539 nautical mile
	= 0.621 statute mile
	= 1,093.6 yards
	= 3,280.8 feet
	= 1,000 meters
Knot	= 1 nautical mile per hour
	= 1.15 miles per hour
Mile (nautical)	= length of 1 minute of arc at the equator
	= 1.1516 statute miles
	= 6,080.2 feet
	= 1.853 kilometers
Mile (statute)	= 0.868 nautical mile
	= 1,760 yards
	= 5,280 feet
	= 1.609 kilometers
	= 1,609.35 meters
Month (lunar)	= 29 days, 12 hours, 44 minutes
Temperature	

The following equation can be used to convert the temperature of one scale to that of the other by substituting for the terms:

$$\frac{5}{9} = \frac{°C - 0}{°F - 32}$$

GLOSSARY

Abyssal Hills   Small, irregularly shaped hills located throughout the ocean basins

Abyssal Plains   Extremely flat portion of the ocean floor usually underlain by sediments

Aquaculture   Raising marine organisms under controlled conditions

Atoll   A ring-shaped reef enclosing a landless lagoon

Authigenic Deposit   Deposits formed in place in the ocean basins, that is, precipitated directly from the seawater

Backwash   The return flow of water to the sea after a wave has broken on shore

Bar   An offshore ridge, usually submerged

Barrier Reef   An elongated reef separated some distance from the landmass by a lagoon

Barycenter   A common center of gravity for two astronomical bodies

Bathymetric Chart   A representation of the depth contours of the ocean floor

Bathythermograph   An instrument used to measure ocean temperatures

Beach   The mass of wave-washed sediment which marks the seaward limit of the shore

Benthic Environment   The ocean bottom

Breaker   A wave breaking along the shore

Bioluminescence   Production of light by organisms resulting from a chemical reaction and unaccompanied by heat

Continental Margin   The zone separating the continental landmasses from the sea, composed of the continental shelf, slope, and rise

Continental Rise   The gentle slope that lies seaward of the continental slope

Continental Shelf   A shallow terrace adjacent to the continent that extends from the low-water line to the continental slope

Continental Slope   A relatively steep slope lying seaward of the continental shelf

Coral Reef   An association of colonial organisms that develop into fringing reefs, barrier reefs, or atolls

Corer   A device to recover sediments intact

Coriolis Effect   An apparent force; the effect on objects moving across the Earth's surface which causes objects to veer to the right in the Northern Hemisphere as a result of the Earth's rotation

Current   Movement of water in the ocean; surface currents are wind-driven; others may be induced by density differences

Current Meter   A device to measure the velocity and direction of ocean currents

Deep Scattering Layer   A mass of marine life which causes the reflection of sound waves; it rises to the surface at night and sinks lower during the day

Density   Mass per unit volume

Detritus   Sediments resulting from weathering and erosion

Discontinuity   A boundary that marks the change in physical properties such as salinity, temperature, or density

Diurnal Change   A daily change, usually referring to the 24-hour 50-minute lunar tidal cycle

Echo Sounding   The use of sound to measure ocean depths by determining the time elapsed from the emission of the signal until its return

Ekman Spiral   A model of the effect on surface water of wind blowing across the ocean

Estuary   A semienclosed coastal body of water in which seawater meets and mixes with water from the land

Fathom   6 feet, or 1.828 meters

Fault Scarp   Clifflike structure resulting from faulting of rocks of the Earth's crust

Fetch   The sea-surface distance over which the wind blows and generates waves

Fjord   A narrow, steep-walled ocean inlet

Fringing Reef   A reef attached to the landmass

Frontal System   The line along which air masses of differing characteristics meet

Grab Sampler   A device used to recover geologic samples

Guyot   A flat-topped seamount

Gyre   The circular flow of oceanic circulation

Halocline   The zone in the ocean in which salinity rapidly changes

Hydrophone   Underwater receiver

Insolation   The solar radiation received by the Earth

Internal Wave   A wave produced by density differences within a body of water

Ion   An electrically charged atom or group of atoms and the form in which atoms usually occur in seawater

Isobath   A contour line representing depth on a bathymetric chart

Isostasy   The ideal condition of balance in the lithosphere

Latent Heat   Heat energy available during changes of state

Magma   Hot, molten matter, primarily silicates, within the Earth's interior

Minimum-Oxygen Layer   The region in which oxygen is relatively low

Nautical Mile   1.18 land miles or statute miles (1,852 meters or 6,076 feet)

Neap Tide   A lower tide than normal which occurs when the moon is at first or third quadrature

Nekton   All those pelagic animals which are capable of swimming independent of current movement

Ocean Basin   The ocean floor bounded by the continental margins

Oceanic Rise   An ocean-bottom province which rises from the ocean floor

Paleomagnetism   The study of the Earth's past magnetic field

Paleontology   The examination and study of ancient fossilized life-forms

Pelagic Sediment   Sea sediments derived from marine processes

Photic Zone   The upper layer of water which receives the rays of the sun and the area in which the photosynthetic process occurs

Phytoplankton   The plant forms of plankton

Plankton   The greatest mass of sea life; generally considered to be minute nonswimmers but often applied as a descriptive term to larger forms of life as well

Plunging Breaker   Waves that tend to curl and break with a crash

Reversing Thermometer   A thermometer used to take fixed temperatures at depth

Salinity   The total quantity of solid material in solution in ocean water expressed in parts per thousand

Seamount   An isolated cone-shaped elevation rising over 1,000 meters from the ocean floor

Seismic Sea Wave   See *Tsunami*

Seismic Shooting   A technique utilized to analyze the strata composing the sea floor

Shoreline   The boundary where land and water meet

Specific Heat   The amount of heat required to raise the temperature of a specific mass of a substance

Spilling Breaker   Waves that break gradually over considerable distances

Spring Tides   High tides that are higher than normal, occurring at full and new moon

Surging Breaker   Waves that do not collapse but surge up onto the beach

Swash   The rush of water onto a beach after a wave breaks

Swell   Regular waves which travel out of the generating area

Terrigenous Sediment   Land-derived sediments

Thermocline   The region of the greatest change in water temperatures

Thermohaline Circulation   Vertical circulation, usually due to surface cooling, which results in a convective flow

Tidal Wave   See *Tsunami*

Tide   A regular rise and fall in sea level resulting from gravitational effects of the moon and sun on the Earth

Topography   The features of the Earth's surface

Trophic Level   A level of the food chain; trophic levels occur in succession

Tsunami   (Erroneously called tidal waves); long-period waves produced by catastrophic undersea activity such as earthquakes and volcanic eruptions; generally no more than a few inches in height in the open sea, but upon reaching a coastline they can reach heights of 100 feet

Turbidity Current   A dense current of water and sediment that flows downslope in response to gravity

Turbulence   An irregular movement of water particles

Upwelling   A movement of water from some depth to the surface

Water Mass   A body of water with relatively stable and uniform physical and chemical characteristics

Wave   A movement of water generally induced at the surface as a result of wind stress; it may also be induced at depth and by other factors

Wave Height   The vertical distance between the crest and trough of two successive waves

Wavelength   A horizontal distance between two similar points on successive waves measured parallel to the direction of travel

Wave Period   The time required for two successive wave crests to pass a given point

Wave Refraction   The change in direction of a wave entering shallow water

Zooplankton   The animal forms of plankton

INDEX